CLIMATE CHANGE
PAST, PRESENT & FUTURE

A Very Short Guide

Warren D. Allmon
Trisha A. Smrecak
Robert M. Ross

Paleontological Research Institution
Ithaca, New York
2010

ISBN 978-0-87710-491-9
Library of Congress catalog number 2010924678

Paleontological Research Institution Special Publication No. 38

1259 Trumansburg Road
Ithaca, New York 14850 U.S. A.
http://www.priweb.org

Cover design and layout by Warren Allmon and Amie Patchen.
Interior design and layout by Paula M. Mikkelsen.

Updates to this book are available on PRI's Global Change Project website, http://www.museumoftheearth.org/outreach.php?page=overview/globalchange.

On the cover (counterclockwise from upper right): Cyclone Catarina, photographed from the International Space Station, 26 March 2004 (National Aeronautics and Space Administration photograph); Stora Enso Pulp and Paper Mill (see Figure 26); Hurricane Rita evacuees from Houston, Texas, 21 September 2005 (photograph by U.S. Federal Emergency Management Agency); the "Blue Marble" image of Earth taken on 7 December 1972 by Apollo 17 astronauts on the way to the Moon at a distance of 29,000 kilometers (photograph by National Aeronautics and Space Administration); woman in Sri Lanka being rescued during monsoon flooding, 01 November 2008 (photograph by "trokilinochchi" via Wikimedia Commons); polar bear on an ice flow (see Figure 46); *Mastodons on South Hill* [Ithaca, New York], by William C. Dilger, 1952, gouache on paper (reproduced with permission of the artist).

CONTENTS

"Before I draw nearer to that stone to which you point," said Scrooge, "answer me one question. Are these the shadows of the things that Will be, or are they shadows of the things that May be only?"

Still the Ghost pointed downward to the grave by which it stood.

"Men's courses will foreshadow certain ends, to which, if persevered in, they must lead," said Scrooge. "But if the courses be departed from, the ends will change. Say it is thus with what you show me!"

The Spirit was immovable as ever.

- Charles Dickens, *A Christmas Carol* (1843)

Figure 1. *Modern Earth from space. Photograph by National Aeronautics and Space Administration.*

PREFACE & ACKNOWLEDGMENTS

There is no shortage of books about climate change. Why one more?

This book is an outgrowth of educational outreach activities focused on the general theme of "global change" developed over the past 15 years at the Paleontological Research Institution (PRI) in Ithaca, New York. These activities began as part of general educational programs in paleontology and Earth science, and accelerated greatly as we designed the exhibits and programs associated with PRI's Museum of the Earth, which opened in 2003. The explosion of interest in global change in general, and climate change in particular, since that time led to a more ambitious set of initiatives at the Institution, which we call the Global Change Project. The Project includes exhibits, K-12 curricula, public lectures and presentations, and teacher professional development and website resources. As we developed these materials, we eventually discovered – to our surprise – that despite the multitude of books and websites available, there was no single hard-copy source to which we could direct interested students, teachers, or members of the general public as a concise, user-friendly handbook on the issues that these diverse consumers want to know about, and that also provides an easy entry into other books and articles on the subject of climate change. We hope that this book will serve these purposes.

It should go without saying that this little book is only an introduction to a huge, complex, and rapidly growing topic. Our aim is to answer basic questions, with enough detail to allow the reader to understand where the answers come from, and to stimulate further learning on the subject. We have tried to be as accurate as possible, but also to write for as broad an audience as possible. In attempting this balance, we have inevitably oversimplified some things. In such cases, we hope that the endnotes and references provided will enable the interested reader to pursue additional details. We have also undoubtedly given some readers more than they want to know. In these cases, we hope that the chapter summaries and boxes will help to clarify the main points.

PRI has a particular perspective on climate change, being an institution with roots in geology and paleontology. The Earth has undergone numerous climatic shifts throughout its 4.6 billion years, some understood in great detail and some still being actively studied, but one thing is common among all of the Earth's climate shifts: every significant change in climate has dramatically affected the organisms living during those times. This geological point of view provides an important but often overlooked context to the current climate change that our planet is undergoing.

We also hope that this book will help to address what we see as one of the biggest problems in the climate change controversy: lack of understanding of the nature of the scientific process and scientific conclusions. A great deal of the "debate" over climate change – especially the part that the public hears about in the popular media – has not been between scientists, but between those who understand how science works and those who don't, or act like they don't. We therefore also hope that, by reading this book, readers will be encouraged to think and learn more about how science works.

Climate change is, of course, not just a scientific issue; it is also a political and economic issue. And this has made it into one of the most difficult and contentious topics that science has ever considered. This book is explicitly *not* about the politics or economics of climate change (references to several good discussions of these aspects are included in the Sources of More Information section at the end of this book). It is about the *science* of climate change. We are very aware, however, that many articles, books, and reports that have made this same disclaimer have been criticized for pushing a political agenda. So we want to be clear at the outset about what we think the relationship is between science and politics, at least in this case. In our view, science is primarily about understanding how the physical world *is*, not about what we *do* with that understanding. Scientific conclusions, therefore, do not in and of themselves necessarily support one policy over another or require that one or another policy be implemented. Science is, in this respect, "separate" from politics.

On the other hand, science is done by human beings, and so is unavoidably affected by politics, economics, emotions, personalities, and other "un-scientific" influences. Scientific conclusions are always provisional; this means that no matter how much evidence appears to support them, all scientific ideas in principle can, and should, be rejected if sufficient reliable contrary evidence is discovered. Given these realities, the appropriate role for science in political discourse is to state as clearly as possible (and as often as necessary) the state of scientific understanding of a particular topic, and with what confidence particular conclusions are held by the majority of knowledgeable experts.

It is the appropriate role of politicians and governmental officials to seek out, listen to, and respect scientists' views, to expect scientists to be clear in their public statements of those views, and then to decide what to do with that information, given the many other competing, non-scientific demands of governing. In other words, if the best science available leads to a conclusion that makes someone uncomfortable, or counters someone's long-held beliefs, preferences, or political convictions, that does not make the science "bad" or incorrect. It just means that people do not always feel, think, or act in accordance with how science tells us that the world really is.

Figuring out what the best course of action is in any particular case is not the business or responsibility of science, but of society, which must take informed responsibility for such obligations. We think that a major reason for much of the controversy about climate change is the repeated failure of our political leaders to understand these distinctions, and we hope that this book can be a small step toward remedying the situation, before it is too late.

We are especially grateful to numerous colleagues for their assistance in developing the programs and products of PRI's Global Change Project, of which this book is a partial introduction and summary. Elizabeth Humbert did much of the research for the Project's website, which was initially designed by Emily Butler and Jon Sessa. For helpful comments on earlier drafts of this manuscript, we are especially grateful to Thomas Cronin and Natalie Mahawold, and also to Sara Auer, Carlyn Buckler, Richard Kissel, Don Duggan-Haas, and Amie Patchen. Our special thanks go to Amie Patchen for assistance with figures and references, and to Paula Mikkelsen for her usual careful attention to design, layout, and printing.

PRI's Global Change Project has been supported financially by the Park Foundation, the National Science Foundation, Cornell's Department of Earth and Atmospheric Sciences, Cornell Cooperative Extension, and the generosity of Sylvester Johnson IV.

1. INTRODUCTION

Climate – the collective conditions of the various elements of weather (such as temperature, precipitation, humidity, etc.) that prevail in a given area over an extended period of time – affects all of us, every day.[1] It influences where we live and how we make a living, how we build our homes and communities, how we get our food and water, what we own and wear and buy and sell, and when and how we live or die. Throughout human history, climate – and changes in climate – have influenced the size of human populations, the occurrence of migration, the waging of wars, and the rise and fall of civilizations.[2]

Climate has thus always been a topic of intense human interest and attention, something to which we have had to respond. Yet never before in human history has the climate of the entire world been an issue of concern. Scientists have recognized since the 19th century, based on simple physical principles, that some gases absorb and retain heat better than others, and that the chemical composition of the atmosphere would likely influence the amount of heat retained in Earth's atmosphere. Since at least the early 1980s, however, respected scientists have been issuing serious warnings that global climate is actually changing, and that human activity is likely responsible for this change. Although initially attracting relatively little public notice, these conclusions became of much wider interest after several years of extreme weather conditions in Europe and North America in the late 1980s and early 1990s. In 1988, the United Nations Environment Programme and the World Meteorological Organization officially established a group of scientists – the Intergovernmental Panel on Climate Change (IPCC) – to study the topic and issue regular reports. By 1990, an increasing number of climate scientists accepted that the world's climate is changing, that humans are the primary cause, and that the results will be immensely challenging. This growing scientific consensus was reflected with increasing confidence in the IPCC reports in 1990, 1995, 2001, and 2007. The 1995 IPCC report laid the groundwork for the Kyoto Protocol, established at a meeting in that Japanese city in 1997. That meeting – the United Nations Framework Convention on Climate Change (UNFCCC) – aimed at creating a consensus for decreasing greenhouse gas emissions.[3] Although most nations in the world signed the Kyoto agreement, the U.S. did not. This, together with the September 11, 2001, terrorist attacks and their aftermath, as well as a very vocal and well-organized effort by individuals and entities skeptical of global climate change (mainly in the U.S.), dampened subsequent coordinated international efforts to address global climate change.

Public awareness of the topic again increased, however, in 2006, with rising gasoline prices and the appearance of a book and feature-length documentary film, *An Inconvenient Truth*,[4] by former U.S. Vice President Al Gore. Gore and the IPCC shared the 2007 Nobel Peace Prize for their efforts to bring the topic to greater attention, further increasing public interest and political pressure to address the issue. Several states, for example California, passed their own laws on reducing carbon emissions, and seven northeastern states formed a coalition called the Regional Greenhouse Gas Initiative (RGGI, which currently consists of Connecticut, Delaware, Maine, Maryland, Massachusetts, New Hampshire, New Jersey, New York, Rhode Island, and Vermont) for implementing a mandatory **cap-and-trade** program.[5]

The results of the 2008 U.S. Presidential election brought to power an administration with the view that global climate change was an urgent priority. Another UNFCCC convention in Copenhagen, held in December 2009, brought hopes for a new, signed and ratified, climate treaty. The results, however, were mixed. No legally binding agreements were signed, but leaders – including those of the largest greenhouse-gas-producing nations – agreed to a system of emission verification and on aid to developing countries to deal with the consequences of climate change.[6]

Box 1: Is the Climate Changing?

- Yes. Scientists have several lines of independent evidence that support the view that the global climate is changing and warming. This evidence includes global temperatures, ocean temperatures, glacier and ice sheet loss, permafrost coverage loss, and other natural indicators.

- For example, according to the IPCC report issued in 2007,[7] since 1850, average global temperature has risen 0.76°C (1.2°F) (see Appendix 1). This might not seem like much, but when most of North America was covered by a mile of ice (approximately 20,000 years ago), the global temperature was only 7°C (11°F) cooler. We also know that 11 of the last 12 years during the interval from 1995-2006 were among the 12 warmest years since we started recording temperature in the mid-19th century.

- Ocean temperature has also been increasing. Scientists estimate that the oceans have been absorbing approximately 80% of the heat that has been added to our planet over the past 150 years. As a result, even at great depths, the ocean is warmer than it was in 1961.

- Glacial coverage on mountains and polar ice caps has decreased in both hemispheres. Large ice sheets in Greenland and Antarctica have also seen dramatic losses in their surface coverage and overall thicknesses.

- **Permafrost**, or soil that is frozen year-round, is common throughout much of the northern reaches of continents in countries such as Canada and Russia. The temperatures at the top of the permafrost layer have been increasing since 1980, by up to 3°C.

- These lines of evidence, along with many others, have convinced most scientists that climate is changing significantly, and that the globe is warming.

Box 2: Why Do Most Scientists Think that Humans are Causing Much of Current Climate Change?

- Carbon dioxide (CO_2) is a "greenhouse gas" that traps heat in our atmosphere.

- Carbon dioxide is generated in nature through many means, but is also generated through the burning of fossil fuels by humans.

- Concentrations of CO_2 in the atmosphere have increased by nearly 50% since the industrial revolution began in the 1850s.

- Since the late 1980s, Earth's average temperature has been gradually rising in a way that cannot be accounted for by natural variation alone. Climate models that incorporate increasing CO_2 explain this warming trend better than any models based on natural variation alone.

- Models that were developed as early as the 1970s have shown that high CO_2 concentrations equate with warm periods earlier in Earth's history.

- Most significant scientific objections to the hypothesis of CO_2-based climate change have been answered sufficiently for most scientists in the past 30 years.

- There are still uncertainties about how the interconnected pieces of the climate system work, but none of these areas of research are expected to disprove the basic hypothesis of human-induced climate change.

The end of 2009 brought the U.S. Environmental Protection Agency (EPA) decision to declare CO_2 an air pollutant under the Clean Air Act and to regulate it accordingly. Although a bill was passed by the U.S. Senate in Fall 2009 that would explicitly aim to reduce greenhouse gas emissions by as much as 50% by 2050, at the time of this writing (February 2010), the chances that such a bill will actually be passed by both houses of Congress any time soon appear remote.

As this short summary shows, the subject of global climate change is not just scientific, but also economic and social, and therefore unavoidably political. Almost all coverage of this subject in the popular media (where most people encounter it) includes some consideration

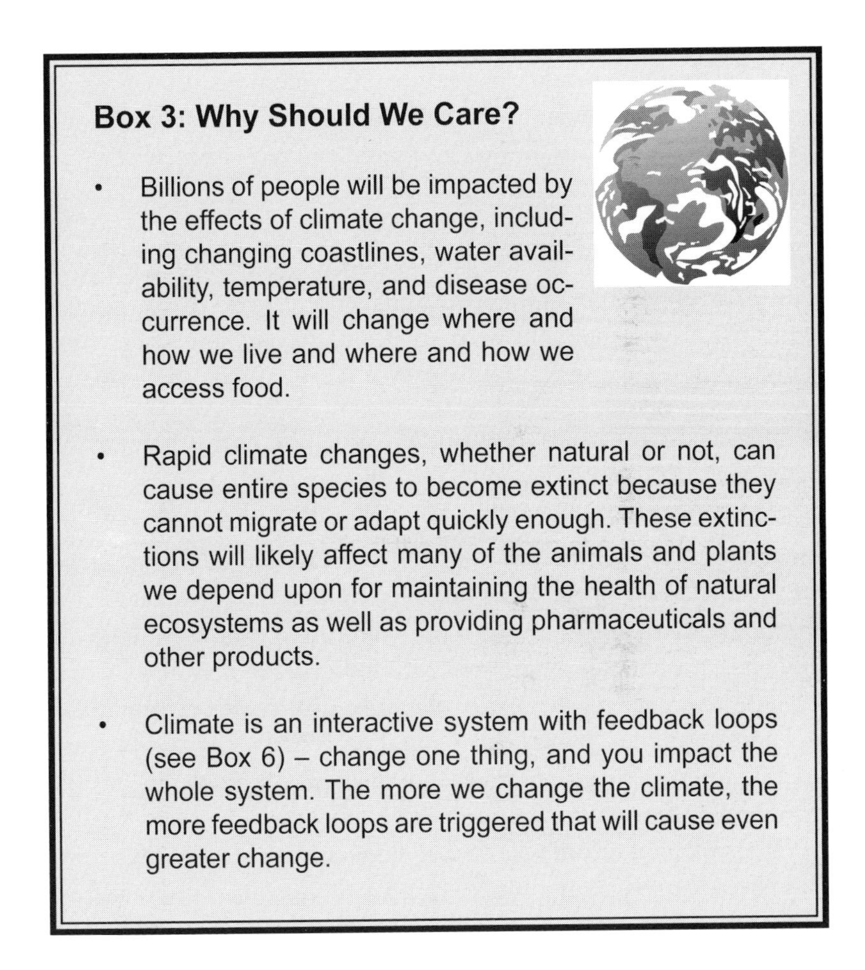

Box 3: Why Should We Care?

- Billions of people will be impacted by the effects of climate change, including changing coastlines, water availability, temperature, and disease occurrence. It will change where and how we live and where and how we access food.

- Rapid climate changes, whether natural or not, can cause entire species to become extinct because they cannot migrate or adapt quickly enough. These extinctions will likely affect many of the animals and plants we depend upon for maintaining the health of natural ecosystems as well as providing pharmaceuticals and other products.

- Climate is an interactive system with feedback loops (see Box 6) – change one thing, and you impact the whole system. The more we change the climate, the more feedback loops are triggered that will cause even greater change.

of its social, economic, or political implications. Thus, although this book focuses on the scientific basis for conclusions about global climate change, it must also address how scientists have reached, presented, debated, and defended these conclusions to the public.

Specifically, this book tries to answer three common questions about climate change (Boxes 1-3):

(1) Is the Earth's climate changing now and if so, in what direction and by how much?
(2) What are the causes of these changes?
(3) How will these changes affect human beings?

We focus particularly on how an understanding of Earth science – including especially the history of the Earth and its life – has led scientists to their current answers to these questions. Our primary goal is to briefly summarize the available data and their most widely accepted interpretations, but we also try to emphasize *how* science has come to these interpretations and *why* they are more accepted by most scientists than alternative views. We hope that readers will take away some sense not just of the details of global climate change, but also of the critical importance of the nature of science in general, how science works the way it does, and what scientific certainty and uncertainty really mean.

The essential message of this book is that ***the great majority of qualified scientists currently accept that global climate change is occurring, that humans (primarily through emissions from burning fossil fuels) are the cause, and that the results will be negative for the well being of the great majority of humans*** (Boxes 1-3). The changes that are occurring are more than just "global warming," although that is among the most significant changes; they also include sea-level rise, changes in weather patterns, disruptions of agriculture and water supplies, damage to natural ecosystems and biodiversity, ocean acidification, and increases in human health problems.

The geological record tells us that climates have been warm before, and have changed many times during Earth's history. But it also tells us that these changes in Earth's past were usually much more gradual – occurring over thousands or millions of years – than many of the changes likely to occur in the very near future. This does not mean that natural climate change has never occurred rapidly – it has. But the geological record also tells us that when very rapid environmental change has occurred, it was detrimental to many organisms who were thriving before the change took place. Indeed, some periods of rapid climate change are associated with very large extinction events. The geological record, furthermore, tells us that our species – *Homo sapiens* – has not experienced anything like what we are about to experience if current trends continue.

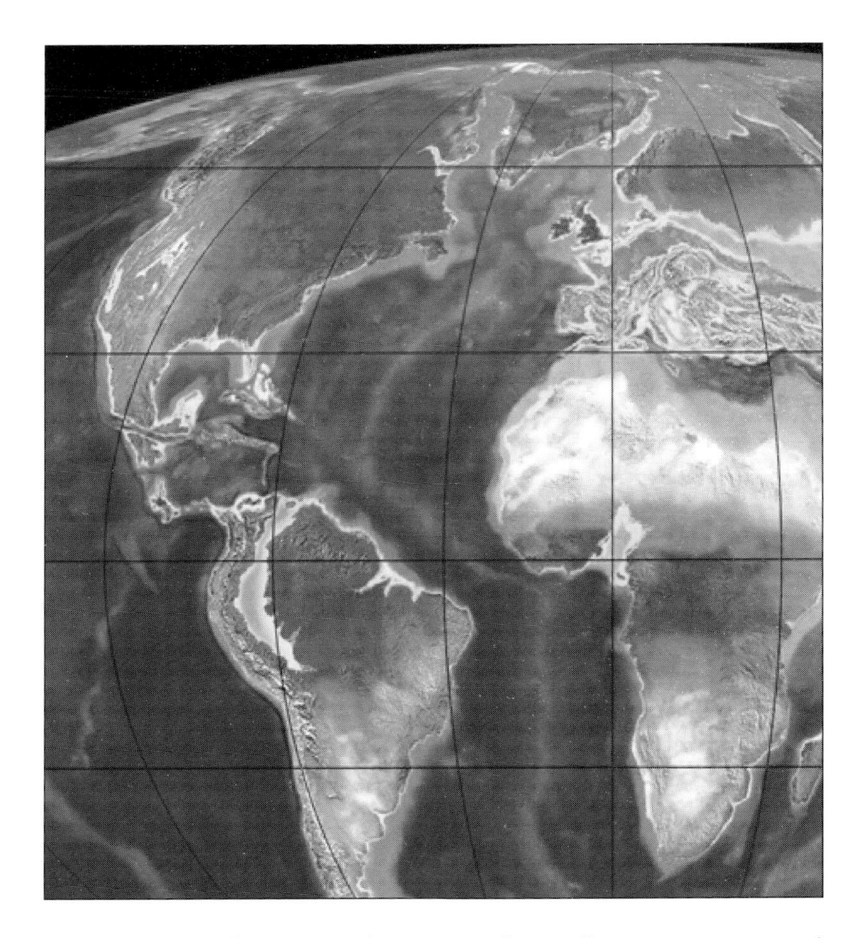

Figure 2. *Earth in the Eocene Epoch, approximately 50 million years ago, was a much warmer place, without polar ice caps or glaciers, and much higher sea levels. Knowledge of past climates helps us understand the present, and predict the future. Graphic by Ron Blakey via Wikimedia Commons.*

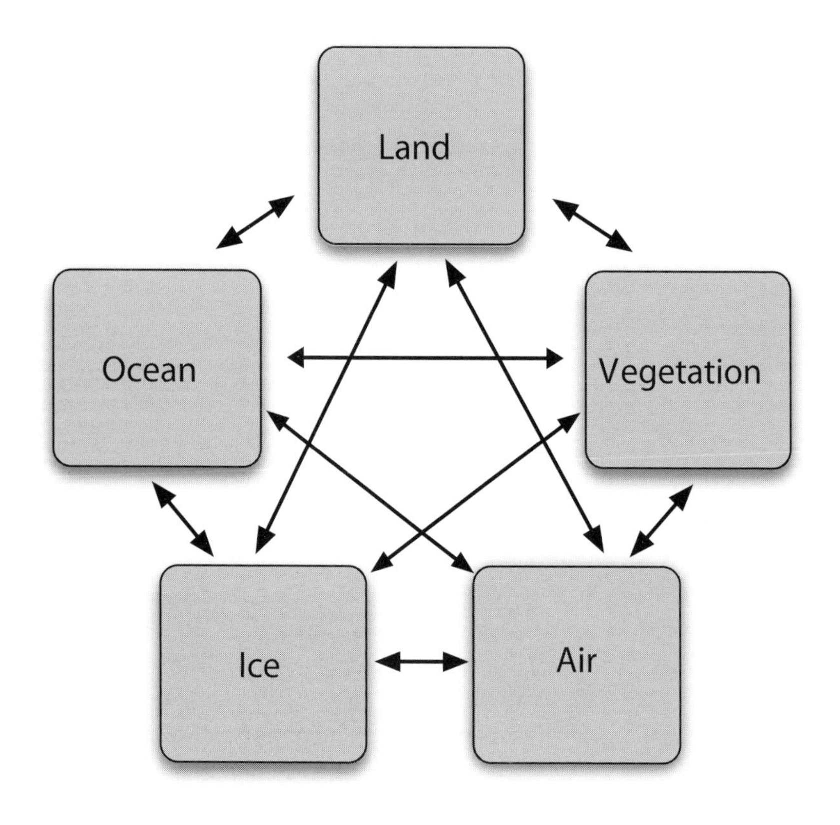

Figure 3. *The climate system. Climate is comprised of multiple, interconnected relationships among land, ocean, air, vegetation, and ice. Each is complex in isolation, but combining changes in multiple components results in hard-to-predict effects on the weather at any given point in space and time.*

2. CLIMATE BASICS

2.1 Climate is a System

Climate is the broad composite of the average weather conditions of a geographic region, as they exist over (at least) decades, measured by temperature, amount of rainfall or snowfall, snow and ice cover, wind direction and strength, and other factors.[8] Fluctuations in these conditions that last hours, days, weeks, or months are called **weather**.

Life exists on Earth today because the climate is favorable for it to do so. In particular, the average surface temperature on Earth allows for abundant liquid water, which is essential for life as we know it. Temperature on our planet is controlled, in part, by the fact that we have an atmosphere of gases surrounding us; these act to hold in some of the energy that we receive from the Sun in the form of light, to keep the Earth warm. Our two closest neighbor planets, Mars and Venus, have very different atmospheres from Earth, and hence, very different surface temperatures. Venus, our neighbor planet closer to the Sun, is thought to have been very similar to Earth when the solar system was born, but today it is very different. It probably experienced a "runaway greenhouse effect," meaning that more and more greenhouse gases were added to the atmosphere without removal, and temperatures on Venus now can exceed 850° Fahrenheit. The atmosphere surrounding Venus is 96% carbon dioxide.[9] Mars, further than Earth from the Sun, similarly has an atmosphere of 95% carbon dioxide, but the total atmosphere is much less dense than that on Earth (Box 4). With such a thin "blanket," little total heat is trapped, thus

Box 4: Why All the Fuss About Carbon Dioxide?

- Carbon dioxide (CO_2) is a molecule made of one atom of carbon and two atoms of oxygen. At temperatures common on Earth, CO_2 usually takes the form of a gas. It is a natural component of Earth's atmosphere, exhaled by animals and used by plants in the process of photosynthesis, so concentrations of CO_2 in the atmosphere vary seasonally as a result of deciduous plants in the northern hemisphere that absorb more CO_2 in the summer months.

- CO_2 is part of the **carbon cycle** (see Figure 10), which includes carbon found in living things, the atmosphere, oceans, and in Earth's crust – in limestone, and in oil, natural gas, and coal deposits. Because carbon usually remains in the ocean and the crust for a long time, these are called "carbon sinks." The carbon cycle is generally in equilibrium, with approximately as much carbon being released into the atmosphere each year as is absorbed by sinks worldwide.

- Some large forests, such as the Amazon rainforest, are shrinking because of human activities, such as deforestation, reducing the size of these sinks.

- Human burning of fossil fuels, such as oil, natural gas, and coal, releases the CO_2 stored in the ancient organic matter. Because these sinks have been long-term repositories for CO_2, their rapid release into the atmosphere is not immediately balanced by any other carbon sink.

- This additional emission of CO_2 into the environment, and the reduction of some carbon sinks, tips the equilibrium of the carbon cycle, so that each year the concentration of carbon dioxide in the atmosphere grows by about 1%. This is believed to be the primary cause of current climate change.

the surface temperature on Mars is -53°C.[10] The relationship between Venus, Earth, and Mars is an example of what has been called the **Goldilocks Principle** – the temperature on Earth is not too hot and not too cold, but "just right" for life to exist.

Climate is a **system** (Figure 3, Box 5). This means that it has parts that interact with each other to create properties that might not exist or be evident in the individual parts examined separately. The Earth's climate is a *complex* system, meaning that it has many parts with many interactions. Complexity reduces (or at least makes more difficult) the predictability of the overall behavior of the system.

The Earth's **climate system** consists of air, water, land, and life (or, as they are often called, the **atmosphere**, **hydrosphere**, **geosphere**, and **biosphere**). Phenomena outside of the Earth (mainly the Sun but also cosmic dust and meteorite impacts) also affect its climate. All of these components interact over time to create the climate conditions that we observe.

The **atmosphere** – the blanket of gas surrounding the Earth (commonly called "air") – is where most of what we think of as weather and climate happen. Other planets, such as Mars and Venus, also have atmospheres, but they are very different from that on Earth. Our atmosphere consists mostly (approximately 80%) of nitrogen, with oxygen making up most of the rest. Other gases exist in much smaller quantities (Table 1). Despite their small quantities in terms of percentage of the atmosphere, some of these other gases – such as water vapor, carbon dioxide, methane, and ozone – have important impacts on Earth's climate system.

The **hydrosphere** includes all the liquid and frozen water at the Earth's surface. The oceans contain approximately 97% of the water on Earth. Because water holds heat for longer than land, the oceans play a very important role in storing and circulating heat around the globe. The currents in the oceans, in fact, are driven primarily by temperature. The surface of the ocean receives heat from the Sun, and this warmer water is less dense than colder water. Therefore, it sits on the surface of the ocean, above the colder, denser masses of water. Winds push the warm surface water, and this movement carries down to the

Box 5: How Systems Work

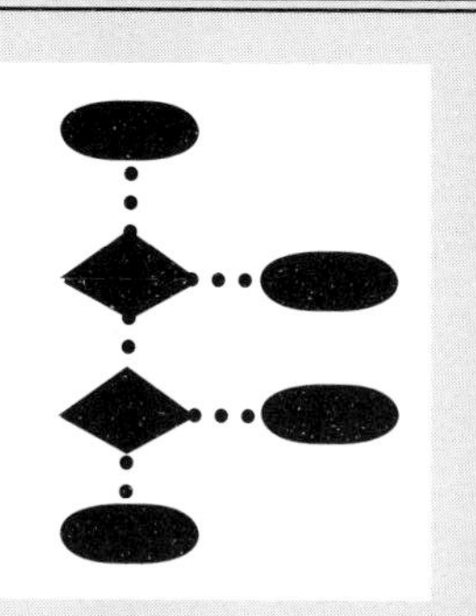

- A **system** is a set of relationships between interrelated factors that have an effect on an overarching unit. So, Earth's climate system is made up of all of the objects and processes that have a global impact on climate. At the simplest level, the components of any system can be analyzed as causes and effects or, as scientists usually say, forcings and responses. The term **forcing** refers to factors that cause change; "responses" are the changes that result. Forcings can produce responses at various rates. These rates can be directly proportional to the magnitude of the forcing (in which case they are called linear; for example, pushing a merry-go-round harder will result in a linear increase in the speed of the merry-go-round), or they can be produced in some more complex, non-linear pattern. A forcing might induce a response immediately after it is applied, or only after some period of time has passed (lag). For instance, when grass seed is planted, there is a lag before your yard is covered by grass, and that amount of time is determined by rainfall, exposure to the sun, and other factors. Forcing can be applied at a variety of magnitudes (strong or weak) and durations (long or short).

- A very important component of all systems is **feedback**. This is change caused by changes already occurring, either by amplifying them (**positive feedback**) or suppressing them (**negative feedback**). An example of a positive feedback cycle is rolling a snowball down a hill. If you make a small snowball and set it on the top of a hill, it begins rolling and collects snow as it rolls. When it collects snow, the snowball becomes heavier, overcoming some of the effects of friction as it rolls, thus making the snowball roll faster down the hill. This causes more snow to collect on the snowball faster, which increases the speed of the snowball (because of the changing snowball mass and frictional forces). Thresholds are the point beyond which the system will change dramatically. There is a point at which the snowball has reached its top speed and cannot go any faster; it has reached its speed threshold. At that point, it maintains its speed until it has reached the bottom of the hill. The threshold temperature for water is 0°C (32°F), below which water will freeze.

layers below, contributing to their motion. In the tropics, most of the warm water at the surface is pushed by wind to the centers of large rotating masses of water called gyres (Figure 4), but some of it also moves toward the poles. When warm water approaches the poles, its temperature drops, it becomes denser, and it begins to sink. It then begins to move back toward the equator, sliding underneath the warmer and less dense surface waters. This is the primary driver of deep ocean circulation.

Table 1. Composition of Earth's atmosphere.

Gas	% in Atmosphere
Nitrogen	77.769001
Oxygen	20.861502
Argon	0.930232
Water vapor	0.398386
Carbon dioxide	0.038145
Neon	0.001811
Helium	0.000522
Methane	0.000174
Krypton	0.000114
Hydrogen	0.000055
Nitrous oxide	0.000030
Carbon monoxide	0.000010
Xenon	0.000009
Ozone	0.000007
Nitrogen dioxide	0.000002
Iodine	0.000001

The movement of ocean currents, carrying heat energy to different parts of the globe and transferring energy to the atmosphere, plays an extremely big role in global climate. Therefore, the con-

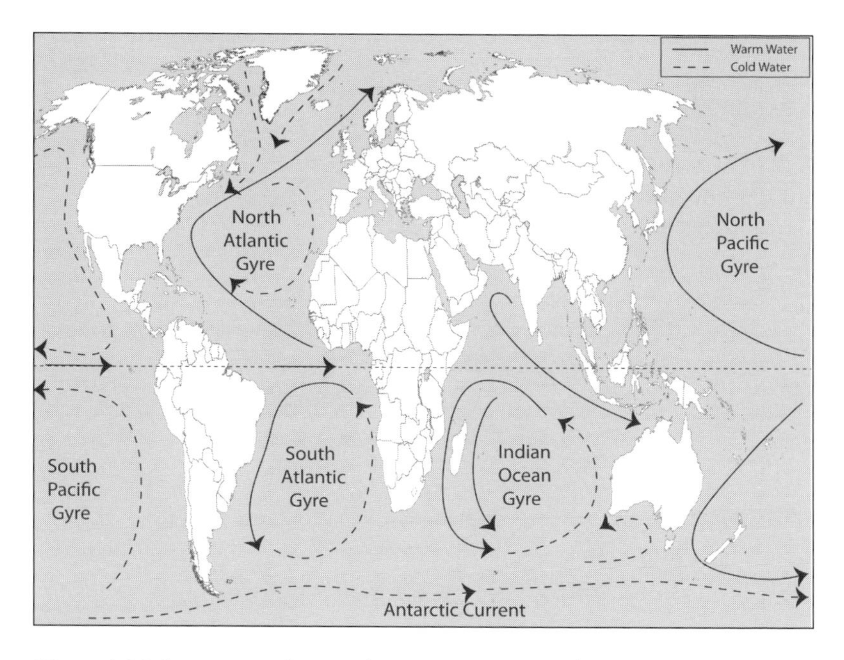

Figure 4. *Modern ocean surface circulation. Ocean currents play a huge role in transporting heat energy from equatorial regions to temperate and polar regions. Surface circulation of a relatively thin layer of water water is driven the wind and by the Coriolis force, an effect of rotation of the Earth, which drives gyres in the Atlantic and Pacific Ocean. Subsurface circulation, which is not shown, is driven by cold salty water sinks near the poles, especially in the North Atlantic.*

figuration of the continents, around which the ocean currents flow, also plays a big role in their respective regional climates.

At higher latitudes, such as around northern Europe, when the warm water begins to cool, its heat is lost to the atmosphere, contributing significantly to warming the air. For instance, the average yearly temperature in London is 14°C (57°F). Land near water in these regions is therefore usually warmer than land far from the coast. At the same latitude across the Atlantic in Calgary, Alberta, Canada, the average yearly temperature is 4°C (39°F). This is because the Gulf Stream carries warm water from near the equator in the Atlantic Ocean northeast to London, but Calgary is in the middle of North America, far away from the moderating influence of ocean currents.

Ice at the Earth's surface includes **sea ice** and **glaciers**, which altogether hold approximately 2% of the water on Earth. Scientists refer to the system as the **cryosphere**. Sea ice forms when seawater freezes at -1.9°C (29°F), which is lower than "freezing" (0°C or 32°F) because of the salt content. Like all ice, frozen seawater is less dense than liquid water, and floats atop it.[11] Sea ice acts as a barrier that prevents the ocean from interacting with the atmosphere. When ice is present, heat from the ocean is not lost to the atmosphere, and the water can remain much warmer than the air. Glacial ice occurs as mountain glaciers or continental ice sheets. Mountain glaciers can occur anywhere in the world, but in the tropics they cannot form below 5 kilometers (16,404 feet) altitude, where it is too warm.

There are currently two continental ice sheets on Earth, covering most of Greenland and Antarctica. These large continental glaciers also lock up great quantites of water that would otherwise be in the ocean, and thereby lowering sea level. If these ice sheets were to melt entirely, global sea level could rise as much as 70 meters (approximately 230 feet). Ice is not only a result of cooler climates; it affects climate itself through its albedo. **Albedo** is the reflectivity of a surface; high albedo means that a surface is very reflective of light energy, and low albedo means that it absorbs light energy as heat. Ice has high albedo; it reflects back lots of sunlight into the atmosphere, cooling the surface. Continental glaciers can be thousands of feet high, and can therefore also actually block or redirect air flow, causing warm air to deflect away from the area covered by the ice sheet, and preventing or slowing the warming process.

The **geosphere** is the solid Earth, from the surface to the core. We might not often think of rocks as being connected to weather, but they very much are, especially over long stretches of time. The solid Earth affects the climate in many ways. Volcanic eruptions can put large amounts of gas and particles into the atmosphere, which can affect how much of the Sun's heat reaches the surface and how much of that is retained. The different land surfaces have varying albedos, variously absorbing and reflecting energy from the Sun. Sediments and rocks hold a large amount of the Earth's carbon, which ultimately affects the atmosphere (see Section 2.3 for more discussion.)

The **biosphere** is all of the life on Earth. Life on Earth is more than just a green layer sitting passively on the surface of a rocky ball; life is an integral part of the geology of the planet. Living things have enormous effects on many geological processes. For example, soil is a byproduct of life; without organic matter, soil would be no more than rock dust. Life also profoundly affects the atmosphere. It is only because of the photosynthetic activity of green plants, along with small organisms like protists and bacteria, that the Earth's atmosphere contains so much oxygen. These organisms can also act as sinks for the carbon that they contain when they die and are buried in rock.

A wide range of organisms help to cycle carbon back to the atmosphere. Plants stabilize the land and limit physical erosion from wind and water (yet they simultaneously contribute to chemical breakdown of rocks by changing the acidity of the soil). Animals (other than humans) alter the landscape in a wide variety of ways, from churning up seafloors and soils to building major structures like coral reefs, beaver ponds, and termite mounds. The remains of dead plants, animals, and microbes form vast deposits of sediment that become layers of rock in the Earth's crust. All of the coal and most of the limestone in the

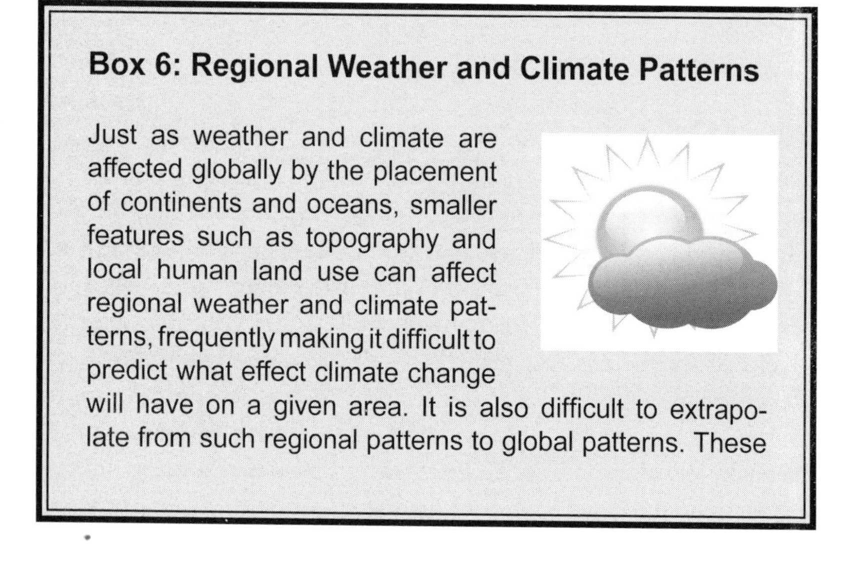

Box 6: Regional Weather and Climate Patterns

Just as weather and climate are affected globally by the placement of continents and oceans, smaller features such as topography and local human land use can affect regional weather and climate patterns, frequently making it difficult to predict what effect climate change will have on a given area. It is also difficult to extrapolate from such regional patterns to global patterns. These regional features create regional and local effects, such as heat islands, rain shadows, and lake effect.

- **Heat islands** occur in urban areas, with the result that such areas are often warmer than nearby rural areas. The building materials used to create metropolitan structures are darker in color and retain heat, therefore the effect is more apparent at night, when the retained heat is radiated back out. This causes some urban areas to be up to 2 or 3°C warmer than their rural counterparts.

- **Rain shadows** refer to the areas adjacent to a mountain range that receive little rain. The mountains separate the area in a rain shadow from a significant water source, like an ocean. As warm air moves over the ocean it collects water in the form of water vapor, then runs into the mountains where it is forced upward. There, the air cools and water vapor condenses to form clouds and rain. The rain falls on the side of the mountains with the water source, leaving the opposite side of the mountains and adjacent land very dry, in a rain shadow (Figure 5).

- **Lake effect** refers to a type of snowfall pattern in which cold air flows over the warmer water of a large lake. Clouds build up over the lake, they get carried to shore by winds, and deposit snow on land in the path of the winds. Since water retains heat better than land, water is slower to cool in the evenings, creating wind patterns that are unique to the areas around the lakes. Therefore, regions along the shores of large lakes, such as central upstate New York or Michigan can have significantly different weather than areas further from the lakes.

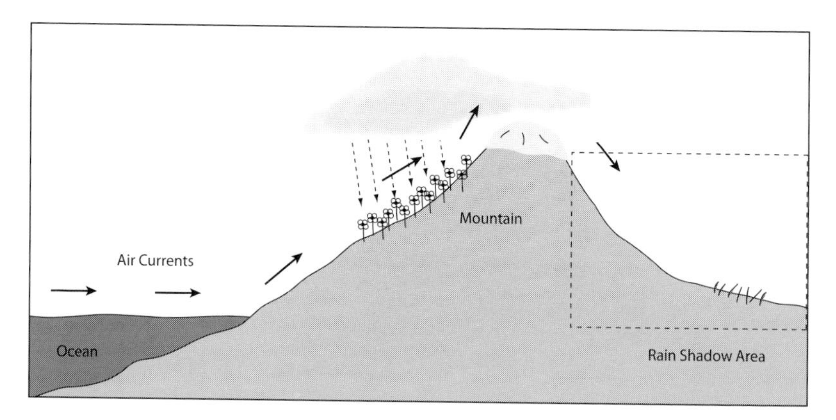

Figure 5. *Rain shadow. As air passes over a water source, it collects moisture. Warmer air collects more moisture than cooler air. Moisture gathers and, along with dust particles in the air, forms rain clouds. As air rises and cools, the moisture precipitates out on the side of the mountainous area closest to the water source. By the time air has risen high enough to pass over the mountains, it has lost most or all of its moisture, thus these areas receive very little precipitation; this is called the "Rain Shadow" effect (see Box 5).*

world, for example, was formed by the accumulated body parts of once-living things.

2.2 Measuring Climate

The temperatures that we hear or read about in the daily weather report are almost always measurements of air temperature obtained by thermometers in particular locations close to the ground (referred to as "near-surface") at particular moments. Such individual measurements can be averaged to produce assessments of temperatures over some geographic area or a length of time. These temperature averages vary on different scales; that is, they vary between different sets of extremes depending upon the length of time or size of the area being considered. Temperatures at a particular spot on the Earth's surface vary by time of day and time of year. Each year is also slightly different, and within the lifetime of an individual human, even larger-scale patterns can become evident; for example, there might be years of higher or lower temperatures or precipitation. In the 1930s, for instance, a severe drought caused the famous "Dustbowl" conditions in the American West and Midwest.

The average global surface temperature (the average of near-surface air temperature over land and sea surface temperature) on Earth during the 20th century was approximately 4.9°C (40.8°F) (see Appendix 1).[12] Many factors affect local and regional temperature patterns, such as land use, topography, and proximity to bodies of water (Box 6).

2.3 How the Climate System Works

The Earth's surface temperature is controlled mainly by the input of heat from the Sun, but this is complicated by the behavior of different components of solar radiation (Figure 6). Sunlight consists of radiation at a variety of wavelengths.[13] The shorter wavelengths pass through the atmosphere largely unobstructed. When sunlight strikes

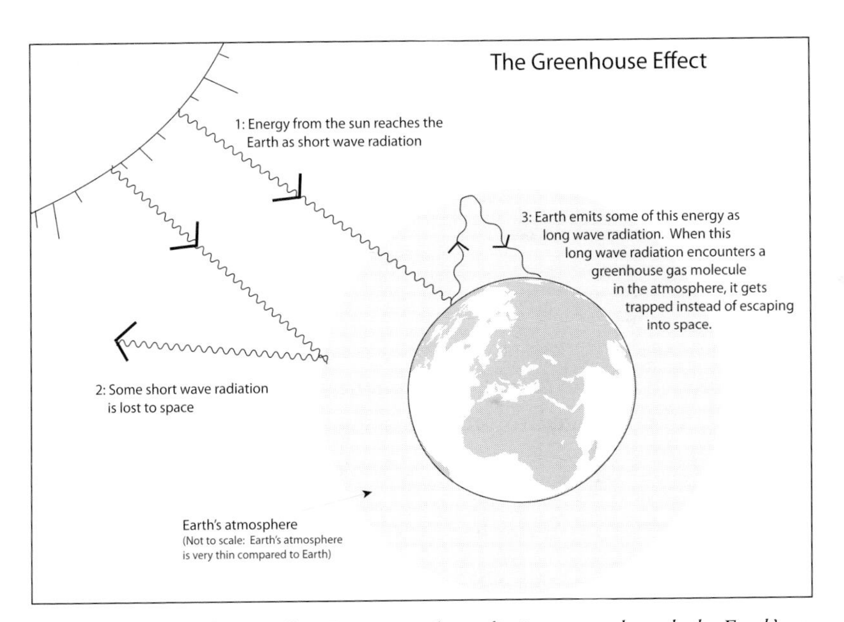

Figure 6. *The Greehouse Effect. Incoming solar radiation comes through the Earth's atmosphere, with some being reflected back before entering. The atmosphere acts somewhat like a blanket, trapping some of the solar energy in the form of heat and keeping Earth warm enough to sustain life. The thicker the blanket, the more solar energy is trapped.*

the Earth's surface, some portion of the radiation is absorbed into the surface (water, rock, vegetation), and some part is reflected back into the atmosphere.

Much of the longer wavelengths of solar radiation is absorbed before they ever reach the surface. This absorption is not due to the atmosphere's dominant gases, nitrogen and oxygen, but to other gases that exist in the atmosphere in much lower concentrations. Because of their more complex geometries, these gas molecules – especially water, carbon dioxide, and methane – absorb radiation more readily than molecular nitrogen and oxygen. They are called **greenhouse gases** (Table 2).

When the shorter-wavelength part of sunlight that has passed through the atmosphere strikes the Earth's surface, it heats the ground, sending longer-wavelength radiation back into the atmosphere, where it is absorbed by greenhouse gases. The atmosphere is heated by this absorption, and itself emits radiation in the form of heat, both upward (toward space) and downward (toward the ground). Thus the surface of the Earth receives radiation from the atmosphere as well as the Sun. Surprisingly, averaged over the entire planet, the Earth's surface receives more radiation from the atmosphere than directly from the Sun.

Greenhouse gases keep the Earth's surface warmer than it would be otherwise, but at the same time, the movement of air dampens this

Table 2. *Common greenhouse gases.*

Water vapor	H_2O
Carbon dioxide	CO_2
Methane	CH_4
Ozone	O_3
Nitrous oxide	N_2O
Chlorofluorocarbons (CFCs)	Composition varies, but commonly include C, Cl, F, and H

warming effect and keeps the surface temperature within the bounds currently experienced on Earth. If the air of the atmosphere were motionless, its greenhouse gases would succeed in raising the average temperature of the Earth's surface to approximately 30°C (85°F), much warmer than we observe. This does not happen because warm air from near the surface rises upward, and is continually replaced by cold air moving downward; these flows of air, called **convection currents**, lower the Earth's surface temperature to an average of 16°C (60°F), while warming the upper reaches of the atmosphere.

Among greenhouse gases, water vapor actually has the greatest capacity to absorb longer-wavelength radiation. In studying changes in

Box 7: Measuring Gases in the Atmosphere

The concentration of a gas in the atmosphere is commonly measured in parts per million (ppm). A value of 1 ppm means that one molecule is present in every million molecules of air. One molecule in a million does not sound like a lot of molecules, but one cubic centimeter of air at the Earth's surface contains approximately 2.7×10^{19} molecules, so a 1 ppm concentration of a gas has 2.7×10^{13} molecules in the same small volume. That's 27 trillion molecules of CO_2 in the space of a sugar cube!

The emission of CO_2 into the atmosphere is commonly expressed in "tons." A single ton (2,000 pounds) of carbon corresponds to 3.67 tons of CO_2 because of the additional weight of the oxygen. To raise the atmospheric concentration of CO_2 by 1 ppm requires 5.9×10^8 (5,900,000,000) tons of CO_2, which is approximately 1 ton of CO_2 per person on Earth. Human burning of fossil fuels, for example, adds approximately 7×10^9 (7 billion or 7,000,000,000) tons of carbon to the atmosphere annually (see Section 5.2).

the Earth's surface temperature over time scales of more than a few weeks, however, more attention is usually given to CO_2 because water vapor concentration in the atmosphere changes much more quickly than does CO_2. For example, a molecule of water vapor, such as might evaporate from the ocean, will remain in the atmosphere for approximately two weeks, whereas an average molecule of CO_2, such as you might exhale, will remain in the atmosphere for several hundred years.[14]

Carbon dioxide concentration in the modern atmosphere varies seasonally over a range of 5-6 ppm (see Box 7). This is because of the activity of forests in the Northern Hemisphere. Forests take in CO_2 (through photosynthesis) in the spring and summer, and release CO_2 (through the slowdown of photosynthesis and the decay of fallen leaves) in the fall and winter. The cycle is reversed in the Southern Hemisphere, but there is much less land area and so fewer forests in the Southern Hemisphere; therefore the Southern Hemisphere effect is much smaller and seasons in the Northern Hemisphere dominate the actual CO_2 cycle.

The average annual concentration of CO_2 in the atmosphere prior to the Industrial Revolution (when large quantities of fossil fuels began to be burned by humans) was approximately 280 ppm.[15] This has been determined by looking at CO_2 trapped in air bubbles in ice sheets (see Section 3.3).[16] Figure 7 shows the global changes in five major greenhouse gases – CO_2, methane, nitrous oxide, CFC-12, and CFC-11. It was not until the 1950s that scientists began to continuously measure the concentration of CO_2 in the atmosphere with high degrees of accuracy and on a regular basis (see Figure 28).

Greenhouse gases are not the only components of the atmosphere that affect global temperature. **Clouds** (masses of tiny water droplets) influence climate in a variety of ways and on a variety of spatial and temporal scales. Clouds can cool the Earth by reflecting sunlight back into space. They can also warm the Earth by reflecting infrared radiation back to Earth.

The amount of water vapor (and thus of clouds) in the atmosphere is sensitive to temperature: the warmer it is, the more water evapo-

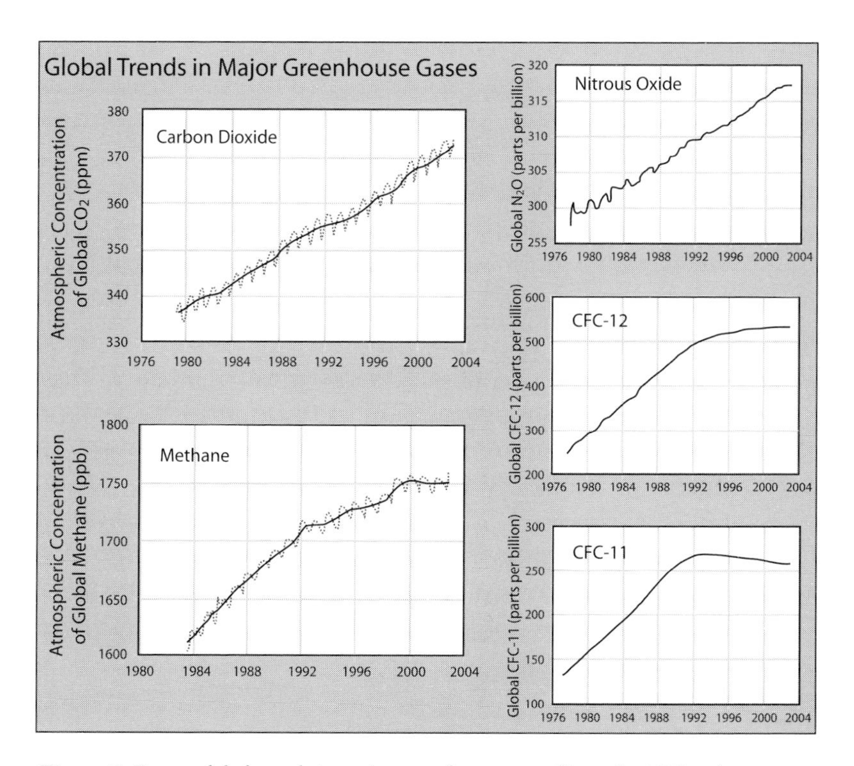

Figure 7. *Recent global trends in major greenhouse gases. Since the 1970s, the amounts of CO_2, methane, and nitrous oxide have all increased in our atmosphere. CFC-11 and CFC-12 are two common types of chlorofluorocarbons that were used in refrigerants and for other industrial purposes. Although they can deplete ozone through the interaction between ultraviolet light and chloride, without the interaction, they are significant greenhouse gases. Regulation of industrial use of CFCs has caused the flattening of the curves seen in the graphs, because there are no longer substantial emissions of CFCs into our atmosphere. Modified after graphs by the National Oceanic and Atmospheric Administration, based on data from IPCC (2007, see endnote 7).*

rates, and the more water that the air can hold – approximately 6% more water vapor for every °C of additional heat.[17] This creates positive feedback in the climate system: the warmer it gets, the more water vapor there will be in the atmosphere, and this will cause still more warming.

Another variable is **aerosols**, which are solid or liquid particles suspended in the air – from volcanic eruptions, storms, or anthropogenic emissions. Aerosols can cool the Earth by both reflecting incoming sunlight and also serving as "seeds," or **condensation nuclei**, for

Box 8:
The Greenhouse Effect

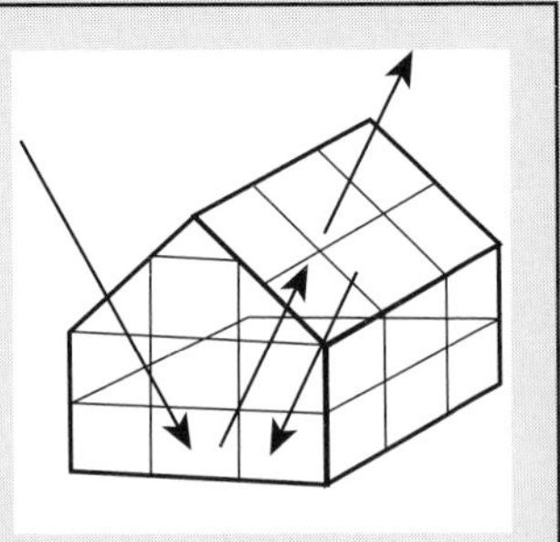

A greenhouse works by letting energy from the Sun indoors through its windows. Those same windows act as a shield from the wind, which would otherwise carry the solar energy away. In the atmosphere, what is commonly referred to as the "greenhouse effect" is much more complex (Figure 6).

Step 1: Earth absorbs energy from the Sun in the form of shortwave radiation, which heats the planet's surface.

Step 2: Earth emits some of this heat in the form of longwave radiation.

Step 3: Some of the longwave radiation being given off by the planet strikes molecules of greenhouse gases in the atmosphere and is absorbed, warming the air.

Step 4: Because of the chemistry of greenhouse gases, longwave radiation is more easily trapped than shortwave radiation. As a result, much of the heat given off by Earth is retained by the atmosphere instead of being allowed to pass through.

The greenhouse metaphor is not a perfect one. Greenhouse windows work to let heat into the building in the form of shortwave radiation – light. They then protect that heat from being dissipated or carried away by winds, locally providing heat to the plants inside. Earth's atmosphere, on the other hand, keeps longwave radiation – heat – from radiating away from Earth and into space.

clouds. The number and size of aerosol particles determines whether the water in clouds condenses into a few large droplets or many small ones, and this strongly affects the amount of sunlight that clouds reflect and the amount of radiation that they absorb. The increased reflection of sunlight into space by aerosols usually outweighs their greenhouse effect (Box 8), mostly because aerosols remain in the atmosphere for only a few weeks.

2.4 Causes of Climate Change

As we ask and answer the question of why climate changes, we must simultaneously consider the temporal *scale* of our discussion, that is, the extent of time over which changes occur (Table 3). Earth has been in existence for 4.6 billion years, and life has been visibly thriving on it, in one form or another, for most of that time. Thus, what has happened in the last 100 years is only a tiny part of the history of Earth and its life and climate. Some causes of climate change have

Table 3. *Some common causes of climate change in Earth's history and their temporal scale.*

Climate Change Cause	Scale of Change
Position in the solar system	billions of years
Heat generated by the Sun	billions of years
Evolution of photosynthesis, other biological impacts	millions to billions of years
CO_2 input from volcanism	millions of years
CO_2 removal from weathering	millions of years
Plate tectonics	millions of years
Shape of Earth's orbit around the Sun (eccentricity)	hundreds of thousands of years
Tilt of Earth's axis relative to the Sun (obliquity)	tens of thousands of years
"Wobble" of Earth's axis (precession)	tens of thousands of years
Strength of Sun's rays based on Earth's tilt	seasonally

tremendous influence, but are only apparent over a million years or more. Others are smaller, but their impacts are seen more readily over shorter time scales, in decades or hundreds of years.

On the scale of millions of years, climates change because of plate tectonic activity. **Plate tectonics**, the mechanism that moves the continents across the globe and forms new ocean floor, has many effects on global climate. Plate tectonic activity, for example, causes volcanism, and extended periods of high volcanic activity can release large amounts of greenhouse gases into the atmosphere. Volcanism also creates new rock, as magma is expelled from the interior of the Earth and cools on the surface. In underwater volcanic activity, new rock can displace ocean water and increase global sea level, which changes the way the oceans distribute heat, and further impacts global climate. For example, the Cretaceous Period, from 145 million to 65 million years ago (see Appendix 2), was a particularly warm period in Earth's history, in part due to the high amounts of greenhouse gas emission from volcanism, and was also a time of higher global sea level.[18]

Plate tectonics also impacts climate on the scale of millions of years due to the changing location of the continents. Climate on land is heavily influenced by ocean currents, so global climate is significantly different when the continents are close together (as in the supercontinent Pangea, which came together approximately 250 million years ago) versus when they are more widely separated, as in modern times. Also, land masses in the equatorial regions have a different impact on climate than continents in higher latitudes because of how heat is distributed from equatorial regions north- and southward around land masses. Therefore, the position of plates over time has had significant impacts on past global climate.

On the scale of hundreds of thousands of years, climates change because of periodic oscillations of the Earth's orbit around the Sun, called **Milankovitch Cycles** (Figure 8; see also Section 2.4.1). These oscillations primarily affect the subtlely varying amount of sunlight received over the course of the year and the distribution of that sunlight across latitudes. Glacial intervals can occur when, in part as a result of these orbital variations, high latitudes receive less summer sunlight, so that their cover of ice and snow does not melt as much.[19]

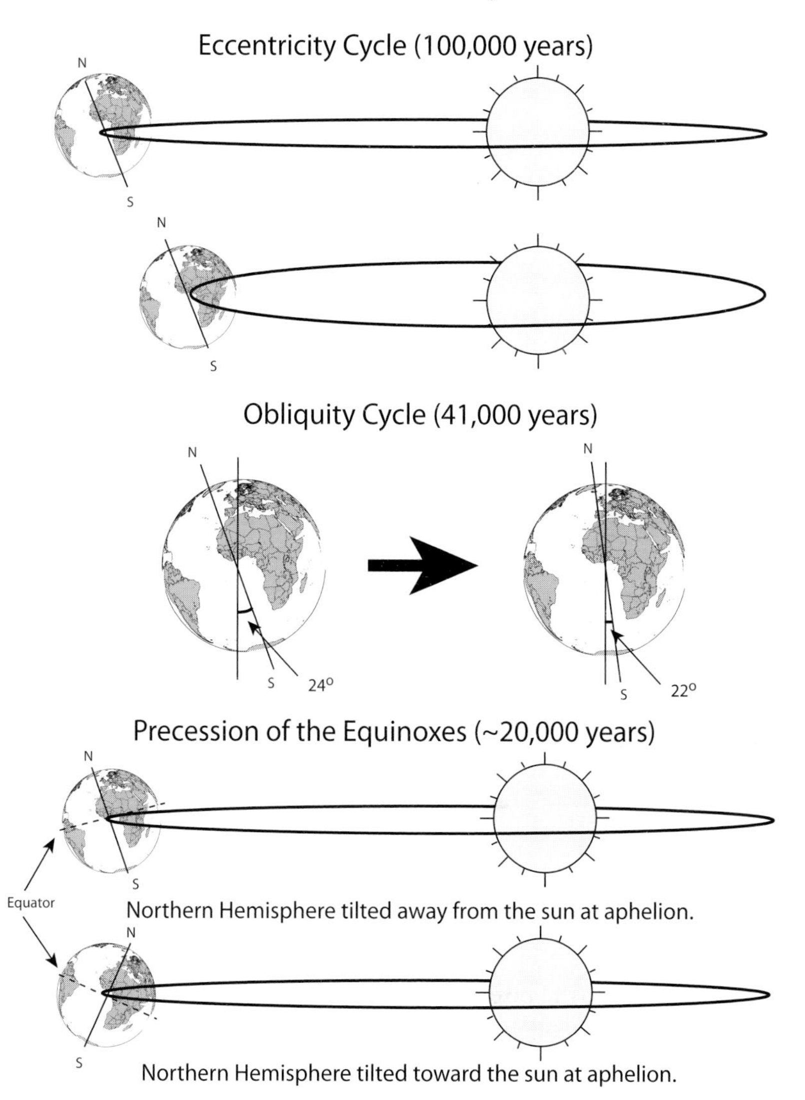

Figure 8. *Milankovitch Cycles. The Earth's orbital variation around the Sun experiences cyclic changes in shape.* ***Eccentricity*** *changes the shape of the orbit on a 100,000-year cycle from a circular to a more elliptical shape.* ***Obliquity*** *is the change of the angle of Earth's axis, which ranges from 22° to 24° from normal, and occurs on a 40,000-year cycle.* ***Precession****, commonly called the "wobble" of Earth's axis, affects the positions in Earth's orbit at which the Northern and Southern Hemispheres experience summer and winter. This changes on an approximately 20,000-year cycle.*

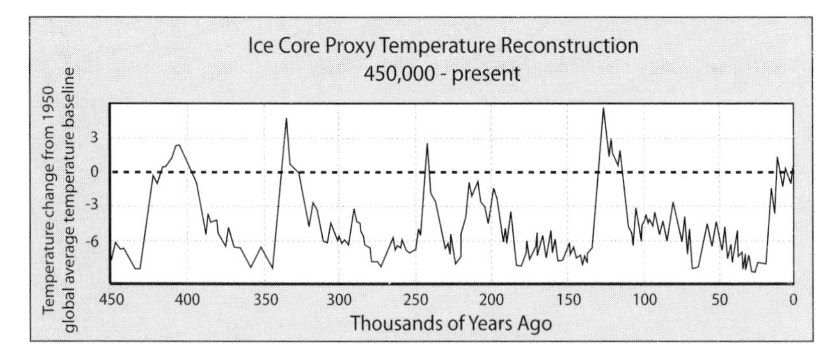

Figure 9. *100,000-year temperature cycles. Ice age temperature changes for the last 450,000 years in this diagram are represented as differences of temperature (in °C) from a modern baseline. These differences are called* ***temperature anomalies*** *(see Appendix 4 for more information). The graph shows abrupt temperature spikes approximately every 100,000 years, each followed by slower cooling. The highest temperatures occurred just after the global climate changed from glacial to interglacial. These temperature changes correlate with changes in the shape of Earth's orbit (due to Milankovitch Cycles). According to this pattern, Earth should now (during this interglacial period) be experiencing slow cooling, not warming. Modified after a graph produced by Robert A. Rohde for Global Warming Art.com.*

Thus, the record of dramatic cooling followed by slow, steady warming (seen in Figure 9) reflects repeated glaciations every 100,000 years or so, caused in part by Milankovitch Cycles.

On the scale of millennia (thousands of years), climates have changed because of cyclic events such as Heinrich events. **Heinrich events** occurred every 7,000-13,000 years and are evidenced by sediment layers on the northern Atlantic Ocean floor, deposited by the melting of huge ice sheets with small rocks and debris contained in them. Scientists believe that these were caused by large icebergs that were released from Canada that, after floating into warmer waters, melted and released large quantities of freshwater. This changed ocean circulation because the large, quick releases of freshwater are less dense than the seawater, and (as we learned earlier) the density of water drives ocean circulation. These large, abrupt releases of freshwater caused a switch from glacial to interglacial types of ocean current patterns.[20]

On the scale of human experience and history (centuries to decades), climates change for a number of reasons. Some are cyclic, and others are the culmination of small changes in topography, land use, etc., that occur in this relatively short span of time. Two examples of changes on this scale are the Younger Dryas event and the Little Ice Age. The **Younger Dryas event** was a 1,200-year interval of colder temperatures that punctuated a warming trend that began approximately 13,000 years ago. Scientists have ascertained that a shift from warming to cooling happened over the course of only a few decades, and brought back glacial climate characteristics such as mountain glaciers in New Zealand and intense windstorms in Asia. One hypothesis suggests that the Younger Dryas was triggered by an ice dam breaking and sending large amounts of freshwater into the northern Atlantic Ocean. Other hypotheses, including one postulating a vast amount of sea ice breaking off of an Arctic ice sheet and floating southward, again sending large amounts of freshwater into the North Atlantic, and even another suggesting a meteor impact, have been offered to explain the Younger Dryas. One thing seems certain – the Younger Dryas is an example of how a single event can reverse or significantly change global climate within a matter of decades.

The **Little Ice Age** occurred between approximately the years 1200 and 1800 CE and followed a time in history called the **Medieval Warm Period**, which peaked approximately 1,000 years ago. The difference in temperature between the Medieval Warm Period, which allowed the Viking people to inhabit Greenland, and the Little Ice Age, which kept Icelandic fishermen frozen in port for up to three months per year from the 1600s through 1930, was only approximately 1°C globally.

Many factors affect weather and climate on the scale of a few years, and some of these can be cyclic or nearly so. One of the most important of these is El Niño.[21] **El Niño** is a climate pattern that occurs across the tropical Pacific Ocean every 3-7 (usually 5) years, characterized by warming ocean surface temperatures and accompanying major shifts in precipitation in the Americas and ocean circulation in the eastern Pacific.[22]

2.4.1 The Sun and Orbital Variations

The Sun is the source of most incoming energy on Earth;[23] it is this solar energy over a given area and time known as **insolation** that controls the energy that drives Earth's climate. Climate models, which will be discussed in Chapter 6, indicate that a relatively small change in the amount of heat retained from the Sun can have a lasting impact on Earth's temperature.

Because nearly all of Earth's atmospheric energy is ultimately derived from the Sun, it makes sense that the planet's position and orientation relative to the Sun would have an effect on climate. The Earth's orbit around the Sun is not a perfect circle, but an ellipse. The distance from the Earth to the Sun changes as the Earth travels its yearly path (see Figure 8). In addition, the axis of the Earth (running from pole to pole) is not vertical with respect to the Sun, but is tilted approximately 23.5°. Earth's tilt is responsible for the seasons, which various parts of the world experience differently. It is summer in the Northern Hemisphere during the part of the year that it is tilted toward the Sun and receives the Sun's rays more directly; conversely, when Earth is on the other side of its orbit and the Southern Hemisphere is tilted toward the Sun, it is summer in the Southern Hemisphere.

Earth's orbit also changes on a longer time scale. **Milankovitch Cycles** (see Figure 8) describe how the position of Earth changes over time in predictable patterns of alternations of the proximity and angle of Earth to the Sun, and therefore have an impact on global climate. These are:

- **Eccentricity** is the change of Earth's orbit from a round orbit to an elliptical one, which occurs on a 100,000-year cycle. When Earth's orbit is more circular, seasons are more subtle.
- **Obliquity** is the tilt of the Earth on its orbital axis, which can range from 22-24° from vertical, and occurs on a roughly 40,000-year cycle. The tilt of the Earth impacts how much insolation is absorbed by the planet at different latitudes.
- **Precession** is commonly called "wobble," because it is the small variation in the direction of Earth's axis as it

points relative to the fixed stars in the solar system. Because of precession, the point in Earth's orbit when the Northern Hemisphere is angled toward the Sun changes over a cycle of 26,000 years.

These three variables interact with each other in ways that can be very complex, but are predictable mathematically. For example, the influence of the shape of the orbit on Earth's climate depends very much on the angle of tilt that Earth is experiencing at the time. The orbital variations described by Milankovitch Cycles are predictable based on the known laws of planetary motion. Confirmation of their climatic effects, however, comes from the geological record, where – in rocks as old as 100 million years – scientists have found indications of environmental conditions (temperature or precipitation) fluctuating in ways similar to what Milankovitch would predict.

The Sun also plays a role in a much more short-term climate cycle through its frequency of solar flares, or **sunspots**, which increase and decrease on an 11-year cycle. When solar flares occur more frequently, the Sun has a larger number of "spots" on its surface and emits more solar energy, which increases the intensity of energy (**irradiance**) that the Earth receives from the Sun. Direct measurements of solar output since 1978 show a rise and fall over the 11-year sunspot cycle, but no overall up- or downward trend in the strength of solar irradiance that might correlate with the temperature increase that Earth has experienced. Similarly, there is no trend in direct measurements of the Sun's ultraviolet output or in cosmic rays. Thus, even though solar irradiance is the primary energy that heats our planet, because sunspots have shown no major directional increases or decreases in their recorded history, they do not appear to be related to the current, directional change in global climate.[24]

2.4.2 The Carbon Cycle

The element carbon plays a crucial role in the way that the Earth works. Because of its particular chemical properties, carbon constitutes the basic building block of living things as well as major constituents of the atmosphere, crust, and oceans. Individual carbon atoms combine

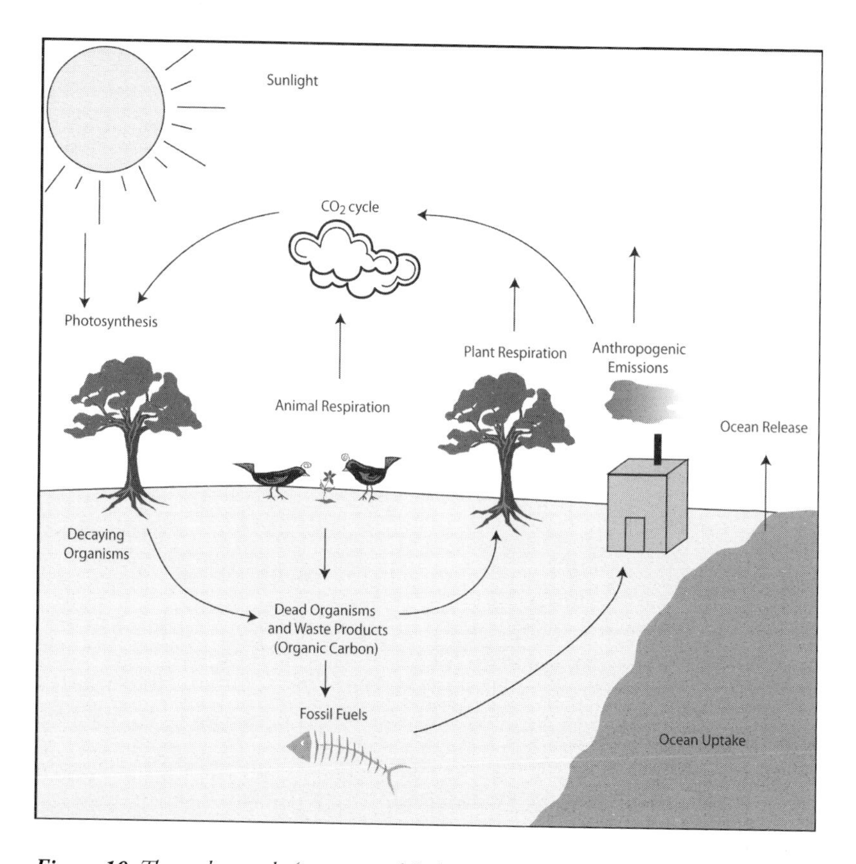

Figure 10. *The carbon cycle (a very simplified view). Every living thing contains carbon. When animals exhale, carbon dioxide (CO_2) is emitted into the atmosphere. It is absorbed by plants through the process of photosynthesis and gets incorporated into their structures. When plants and animals die, the carbon in their bodies gets incorporated into sediments, which might eventually become rocks in the Earth's crust, where it usually remains for millions of years. The extraction and burning of this carbon in the form of fossil fuels emits CO_2 into the atmosphere, and some of it becomes incorporated into carbon sinks like oceans and forests. Omitted from this figure are, e.g., biological processes in the oceans, volcanoes, weathering of rocks, and the formation of limestone.*

with other elements in a variety of ways as they move between these various Earth systems in a series of steps known as the **carbon cycle** (Figure 10). Understanding the role of CO_2 in the Earth's climate starts with understanding how carbon behaves in this cycle.

Carbon dioxide that enters the atmosphere from volcanoes is approximately balanced (in the absence of humans) by removal of CO_2 from the atmosphere by two processes: photosynthesis and **chemical**

Box 9: The Faint Young Sun Paradox

On the scale of billions of years, the Earth's climate has been controlled by the balance between its distance from the Sun and the composition of its atmosphere. If we compare Earth to its planetary neighbors, Mars and Venus, we see what might have happened on Earth, but didn't. Earth's original atmosphere came from volcanism, emitting gases from the planet's interior. The high concentration of CO_2 in this early atmosphere kept the Earth warm when the Sun was younger, and its strength fainter.[25] Some of the first life on Earth put oxygen into the atmosphere by the process of photosynthesis, and drew down CO_2. Lower levels of greenhouse gases in the atmosphere helped to compensate for the brightening Sun, which otherwise would have warmed Earth too much for living organisms, or even liquid water, to be present. Thus, the greenhouse effect allowed life to exist on Earth when it might not have otherwise.

weathering, or the breakdown of rocks at the surface by chemical change. During chemical weathering, water reacts with minerals in rocks and CO_2 from the atmosphere. The CO_2 is thus removed from the air and transferred into other compounds, which eventually become stored in the sediments that accumulate in the ocean and ultimately form sedimentary rocks.

Evolutionary changes in organisms throughout geologic time have had a strong effect on the global carbon cycle. The evolution of organisms capable of photosynthesis, over 3 billion years ago, drew CO_2 out of the atmosphere and created the first significant amounts of atmospheric oxygen. The first appearance of large animals in the early Cambrian, 540 million years ago, and the evolution of land plants in the Devonian, approximately 380 million years ago, accelerated the cycling of carbon and its burial in sediments. It is widely believed that the evolution of land plants led to a significant drop in CO_2 concentrations in the atmosphere and caused the widespread glaciation of the Carboniferous Period (360 to 295 million years ago).

The carbon cycle functions on a variety of time scales. A single atom of carbon that you exhale (as part of a molecule of CO_2) will on average remain in the atmosphere for hundreds of years, before being absorbed by a plant or other photosynthesizing organism. When the plant dies, that carbon atom could in a few weeks or months be taken up by another plant, or oxidize back into CO_2 and re-enter the atmosphere, or it might be buried in the Earth's crust and remain there for millions of years.

2.4.3 Plate Tectonics

The Earth's surface is like a jigsaw puzzle. It is made up of many huge pieces, or plates, which slide around the globe very slowly, at about the rate that your fingernails grow. The continents are embedded in these **tectonic plates** (Figure 11). Where these plates come together or move apart, earthquakes, mountain building, and many other geologic processes can occur. Plate movement is driven by the Earth's internal heat.

Plate movement can significantly affect climate over millions of years, in several ways. The position of a plate on the globe, and of any continents that might be on top of it, is one determinant of whether that continent will experience glaciation or tropical temperatures. For example, if the plate that now holds North America and Greenland were shifted a bit to the north, North America could be covered in a continental ice sheet right now. Instead, only Greenland is covered in ice because it is positioned further north today. Plate movement also affects climate because when two plates come together, volcanoes often result, adding CO_2 to the atmosphere when they erupt. When plates move apart, in a process known as **seafloor spreading**, hot magma is often released directly into the ocean, bringing CO_2 with it.

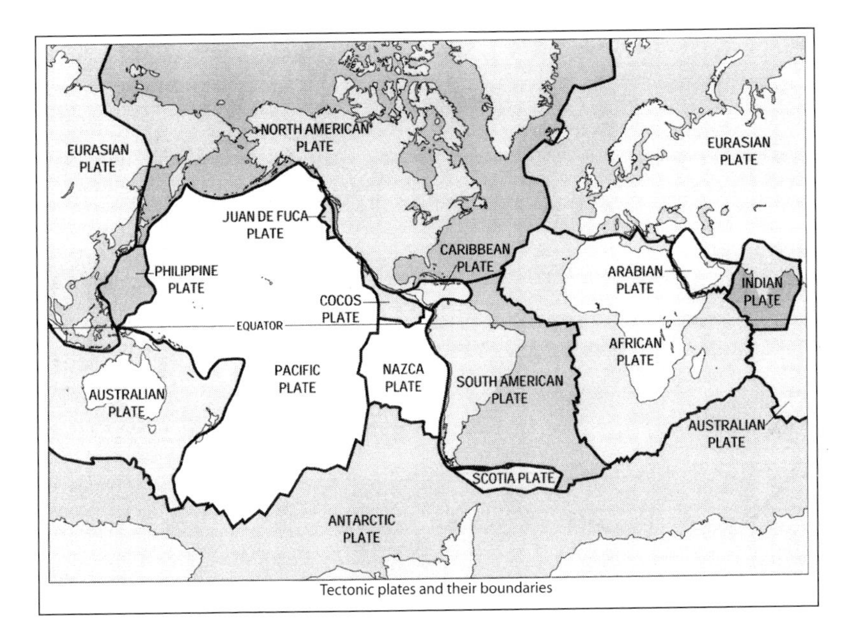

Figure 11. *Plate tectonics. A world map with all of the individual tectonic plate boundaries highlighted. The plates move around like puzzle pieces over the globe. Map by U.S. Geological Survey.*

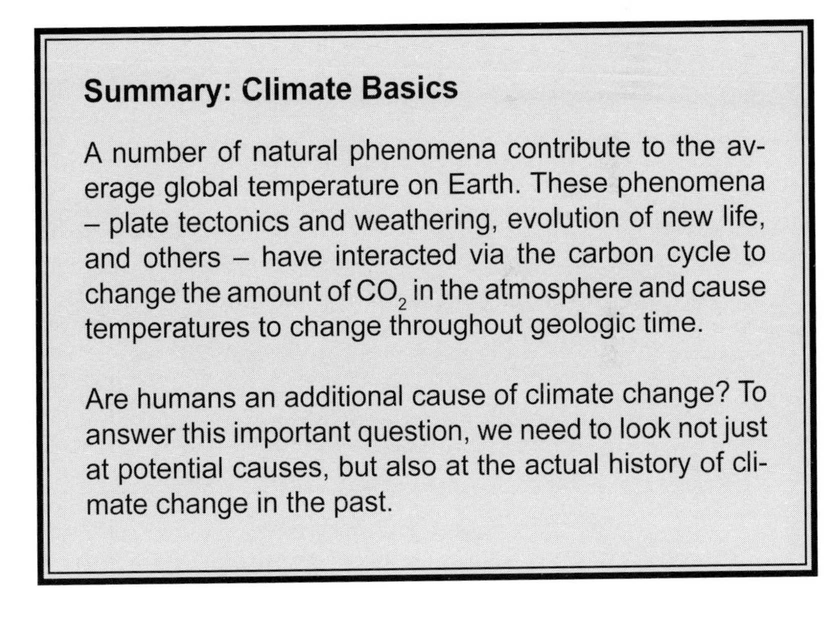

Summary: Climate Basics

A number of natural phenomena contribute to the average global temperature on Earth. These phenomena – plate tectonics and weathering, evolution of new life, and others – have interacted via the carbon cycle to change the amount of CO_2 in the atmosphere and cause temperatures to change throughout geologic time.

Are humans an additional cause of climate change? To answer this important question, we need to look not just at potential causes, but also at the actual history of climate change in the past.

Figure 12. *Bristlecone pines, like this* Pinus longaeva, *in Ancient Bristlecone Pine Forest, White Mountains, California, are among the longest-lived organisms on Earth. Their tree rings have provided scientists with climate data for the past 9,000 years. Photograph by Clinton Steeds via Wikimedia Commons.*

3. CLIMATES OF THE GEOLOGICAL PAST

How do we know what ancient climates were like? How do we know whether they were different from today? To know the average temperature of the world 10,000 years ago, we cannot just look at a thermometer and record the temperature; we need a substitute – a **proxy** – that indirectly recorded that information.

The Earth is a giant, albeit imperfect, recording device. Earth scientists reconstruct ancient climates by using traces left in the rocks and sediments available on the Earth's surface. Even after thousands or millions of years, many of these materials contain information about the environmental conditions that existed at the time that they were laid down at the bottom of a river, lake, swamp, or ocean. This climatic information can be found in unconsolidated sediments (for example, in mud at the bottom of a pond), in rocks, in glacial ice sheets, or even in a living tree or coral colony. Each of these systems records something about the world in which they formed.

3.1 Proxies from Fossils

Fossils – the remains or traces of once-living things preserved in the Earth's crust – can be compared to organisms in modern environments to infer the past environment in which they lived (Figure 13). For example, fossil fish and seashells can reasonably be assumed to have lived in water, even though the place where the fossils were found is now dry land. Fossil reptiles or palm trees found in what are now much cooler, high-latitude locations testify to these areas once having a much warmer climate. Corals are mostly colonial, marine animals that make hard skeletons out of calcium carbonate ($CaCO_3$). Modern corals live mainly in warm, tropical seas. Fossil corals found today in very different environments, such as upstate New York, are therefore indicative of major changes in the climate of the area.

Fossil leaves frequently have characteristic shapes that are, in part, the result of the habitat in which they live. Looking at their shape scientifically with a process called **leaf margin analysis** can help to reconstruct ancient environments and climates (Figure 14). The edges of modern leaves are indicative of their climate and environment;

Figure 13. *Examples of fossil climate proxies. Fossils are used as a proxy to gauge climate change in an area. Top left and right are a fossil palm frond and alligator, respectively, both Eocene Epoch (*ca. *50 million years old), Wyoming (photographs by Paleontological Research Institution). Bottom left shows benthic Foraminifera, found in marine sediments, which can be used as climate proxies; the species are (clockwise from top left)* Ammonia beccarii, Elphidium excavatum clavatum, Buccella frigida, *and* Eggerella advena *(photograph by United States Geological Survey). Bottom right image shows common pollen grains (greatly magnified), including sunflower and lily pollen (photograph by Dartmouth University).*

smooth-edged leaves with narrow, pointed "drip tips" at the ends are common in rainforests where they function to rid the leaves of excess water, whereas toothed leaf edges are more common in temperate environments to preserve water. Scientists measure leaves in modern environments and correlate their size, shape, and edge appearances with the temperature and humidity of the region. That information can then be applied to fossil leaf measurements in ancient environments to calculate approximate temperature and humidity.

Ancient plant pollen and spores (produced by plants such as ferns, lichens, and mosses) can also help us learn about ancient climates. **Palynology** (the study of pollen and spores) uses the fortunate circumstances of these objects being small, abundant, and easily preserved. Due to their tough organic coating, they are commonly preserved in the sand and sediment from places like lakes and rivers, even though trees and leaves are seldom preserved. Pollen can be used just like an entire plant, and if the environmental constraints of that plant are

Figure 14. *Using leaf margins as climate proxies. Plants with leaves with toothed or divided margins (above left) live today in cooler climates, whereas plants with leaves with smooth or entire margins (above right) live in warmer climates. This observation can be used to interpret the climates in which fossil leaves (lower left and right) grew. Photographs by Warren Allmon (top left and right), Alejandra Gandolfo (lower left), and Judith Nagel-Myers (lower right).*

known (by studying it or its descendants living today), the history of climate in the area can be inferred. Pollen and spores have, for example, been used to track how plant communities move north and south during fluctuations between glacial and warmer intervals.

Single-celled organisms, or protists, make up a large proportion of the plankton at the base of oceanic food webs. Some of these protists, especially shelled forms called Foraminifera (Figure 13), are particularly valuable as indicators of past climate conditions, either through analysis of the oxygen isotopes in their fossilized carbonate shells (see Box 10), or by comparing fossil forms to those alive today and inferring that they had similar environmental distributions.

3.2 Proxies from Rocks

Sedimentary rocks are formed through breakdown of other rocks into sediment, which is then transported and deposited by wind or water. When the sediment is compressed or cemented and turned into rock, it retains clues about the environment in which it formed. By observing modern oceans, for example, scientists note that limy sediments and reefs (composed of calcium carbonate) usually accumulate in warm, shallow, clear seawater, and they then use this to conclude that ancient carbonates might have formed in similar environments.

Chemical elements in rocks, and even in some fossils, can also record information about the environment at the time that the rocks were formed. Particularly useful for recreating ancient climates are the different forms (or **isotopes**) of the element oxygen (see Box 10).

3.3 Proxies from Ice Cores

In a few cases, scientists can sample ancient atmospheres directly. Ice sheets and glaciers, which can be hundreds or thousands of feet thick, are formed from snow that has collected each year on the surface, has been compressed by overlying snow and ice for many years, and ultimately recrystallizes into thick glacial ice.[26] Bubbles in this ice can contain air that was trapped when the ice formed. The chemical composition of the air in these bubbles, as well as the frozen water surrounding them, can reveal, for example, the amount of CO_2 in the ancient atmosphere.

An **ice core** (Figure 15) is a large cylinder of ice extracted from an ice sheet or glacier, such as is found in Antarctica, Greenland, or on very high mountains worldwide. To collect ice cores, scientists use a hollow drill that cuts around a central cylinder of ice. Drillers carry out many cycles of lowering the drill, cutting a limited section (usually 4-6 m long), then raising all the equipment to the surface, removing the core section and beginning the process again. Much care is taken to ensure that the core is uncontaminated by modern air and water. The core is stored in an airtight plastic bag as soon as it reaches the surface, and analyzed only in a "clean room" designed to prevent contamination. To keep the ice core from degrading, it is kept well below freezing, usually below -15°C (5°F).

Ice cores record history. They can tell scientists about temperature, ocean volume, rainfall amount, levels of CO_2 and other gases in the atmosphere, solar variability, and sea-surface productivity at the time

Figure 15. *An ice core from Vostok Research Station, Antarctica, at the National Ice Core Laboratory in Denver, Colorado. Photograph by Warren Allmon.*

Box 10: Using Oxygen Isotopes to Determine Past Climates

The different chemical elements, like oxygen, carbon, and hydrogen, that we encounter in the Periodic Table in chemistry class are distinguished by their differing numbers of subatomic particles: each element has a distinct number of protons, and an equal number of electrons. **Isotopes** are variants of elements that have the same numbers of protons and electrons, but differ in the number of neutrons. This means that different isotopes of an element have a slightly different mass.

Higher $^{18}O : ^{16}O$ = colder water/atmosphere

Scientists measure the oxygen isotope ratio in the core and learn what the temperature was like when the calcite formed.

$^{18}O : ^{16}O$ ratio in ocean water

Oxygen isotopes are incorporated into calcite shells of microorganisms in the ratio in which they are present in the water

Sediment Core

Microorganisms die and sink to the bottom of the ocean

Core of rock is taken

Limestone rock forms

The most common isotope of oxygen is oxygen 16, abbreviated ^{16}O, which has 16 neutrons. A small proportion of the oxygen in the universe is oxygen 18 (^{18}O), which has 18 neutrons. Because ^{16}O is lighter than ^{18}O, it behaves differently. For example, it is more easily integrated into water vapor, and so clouds and their associated precipitation contains relatively more ^{16}O than the lake or ocean from which the water evaporated. When this precipitation is stored for a long time in the form of compacted snow in glaciers, as a result of colder climate, the oceans of the world have relatively less ^{16}O in their water than they do in warmer times.

^{16}O is also more easily incorporated into chemical compounds. Many marine organisms make their shells out of calcium carbonate ($CaCO_3$), and need to take oxygen out of the seawater to do this. Therefore, when they build their shells, marine organisms record the proportion of ^{16}O that exists in seawater at the time. Because it is easier for organisms to use ^{16}O than ^{18}O, more ^{18}O is incorporated into their shells only when less ^{16}O is available. Because of the different behavior of the two isotopes of oxygen, shells have a higher proportion of ^{16}O in a warmer climate when the lighter isotope of oxygen is not stored in glaciers. When the shells are preserved as fossils on the sea floor, and then extracted in a **sediment core**, they can be analyzed for their amount of ^{16}O relative to their amount of ^{18}O to estimate ancient temperatures. Scientists commonly use the quantity $\delta^{18}O$ (pronounced "del-18-oh"), which reflects the ratio of ^{18}O to ^{16}O compared to a standard; smaller values mean higher temperatures (*e.g.,* in Figures 16 and 19).

that the ice formed. Scientists can conduct chemical and isotopic studies on the ice itself, but they can also physically look at inclusions in the ice, like wind-blown dust, ash, or radioactive substances that can tell us about the extent of deserts, volcanic eruptions, forest fires, and even meteor impacts. The length of the record is extremely variable. Some cores only record the last few hundred years, whereas the longest core ever taken (from Vostok Research Station, Antarctica; Figure 15) allows study of climate change for over 400,000 years. Compiling information from multiple cores, scientists have now assembled a climate history of almost 800,000 years.[27]

3.4 Using Living Organisms to Determine Past Temperatures

Scientists can also use proxy records stored in living things and their abcient ancestors to recreate climates of the recent past with an amazing degree of precision.

3.4.1 Dendrochronology

Dendrochronology – the study of climate change as recorded by tree ring growth – is an excellent example of how climate researchers get information about climates of the relatively recent past. Trees can live for hundreds of years and in some extraordinary cases, like giant sequoias and bristlecone pines, thousands of years. Each year, a tree adds a layer of growth between the older wood and its bark. This layer, or "ring" as seen in cross section, varies in thickness. A wide tree ring records a good growing season, usually moister and/or warmer, whereas a narrow ring records a poor growing season, usually drier and/or cooler. Especially in environments near the edge of a tree's comfortable living range, such as near the treeline on a mountain, these data can provide highly reliable records of climate patterns.

To know more about climate over an even longer period of time, scientists look at dead trees that are still well preserved. They correlate the dead tree rings with the rings of a living tree whose age is known, looking for overlapping patterns, and can thereby get a longer record of climate through time. An amazing example is seen in the tree-ring chronologies established by looking at bristlecone pines from the southwestern U.S. (see Figure 12). Not only are these trees the longest-lived trees on Earth, they also live in a place where, even when they do die, they remain well preserved for hundreds or thousands of years. By comparing rings of living to dead bristlecone pines, scientists have established a tree ring record of climate in the western U.S. for the past 9,000 years.

There are, however, limitations to dendrochronology. Trees in the temperate zone only record the growing season, so the winter season is not usually visible in their wood. Trees in tropical regions grow year-round and therefore show no obvious annual growth rings, so climate data from equatorial areas is more difficult to obtain.

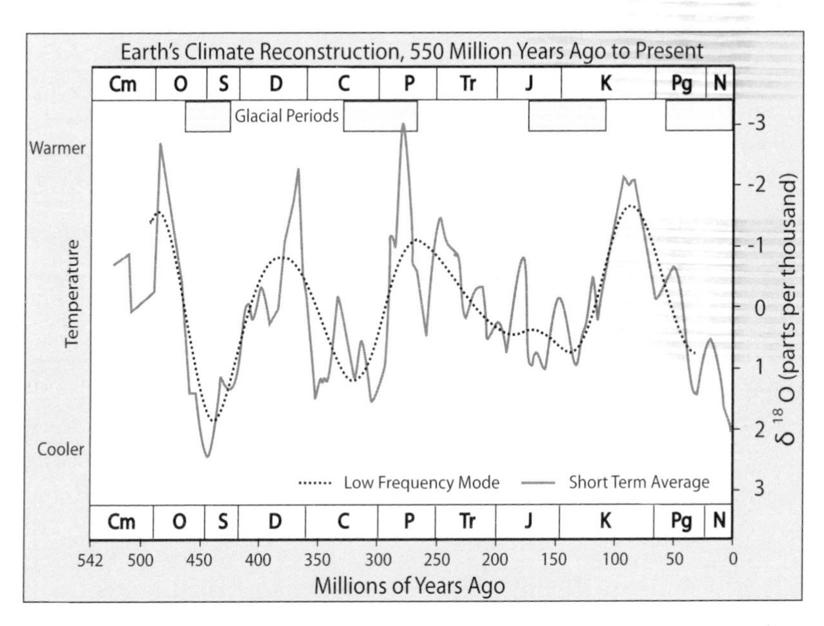

Figure 16. *Earth's average temperature, 542 million years ago to present. Scientists have used geological proxies, including oxygen isotope information (see Box 10), preserved in shell material of some marine fossils, to reconstruct climatic conditions. Because this is proxy data, it only shows relative changes in global temperature, with higher $\delta^{18}O$ values representing cooler global temperatures and lower $\delta^{18}O$ values representing warmer global temperatures. Modified after a graph produced by Robert A. Rohde for Global Warming Art.com.*

3.4.2 Living Coral as a Climate Proxy

Because they build their own calcium carbonate ($CaCO_3$) skeletons, corals (Figure 17) keep a record of climate in a way very similar to trees – by periodic rings of growth in the skeleton. Thicker rings represent better conditions for the coral, whereas thinner rings represent poor conditions. The coral colony grows both in winter and in summer, but the density of the skeleton is quite different due to seasonal changes in ocean temperature, the availability of nutrients, and differences in light. Additionally, the coral rings contain carbon and oxygen isotopes that indicate environmental conditions at the time that that part of the skeleton was secreted.

Using these and other geological proxies, scientists have formed well-supported hypotheses of the history of climates during the last few hundred million years of Earth history. These studies have shown that global climates have been significantly warmer than today, and they have also been significantly cooler. Figure 16 shows a temperature reconstruction of the past 542 million years of Earth's history (roughly since the first appearance of large animals in the fossil record). Although these temperatures are estimates from proxy information, they are based on multiple lines of independent evidence, and this is why scientists have high confidence in them.

Summary: Climates of the Geological Past

The geological record shows that the climate of the Earth has changed dramatically over time. Ancient temperatures are not measured directly, but are estimated using climate **proxies**, including fossils, rocks, and other records. These proxies provide multiple, independent sources of data that not only provide us with reliable information about past climates, but (as we will see) provide methods to check our predictions about future climates.

Figure 17. *A living colony of pillar coral at Discovery Bay, Jamaica. Such colonies can be hundreds of years old and the characteristics of older layers can provide proxy evidence of past environmental conditions. Photograph by Warren Allmon.*

Figure 18. *Landscape reflects changing climate. East fork of the glacially fed Toklat River, Denali National Park, Alaska. As evidenced from its steep river banks and high terraces, the Toklat was a large, flowing river thousands of years ago. Today it is never more than a set of braided streams. Photograph by Clinton Steeds via Wikimedia Commons.*

4. CLIMATES OF THE RECENT PAST

As discussed in the preceding chapter, when trying to reconstruct ancient climates for which no instrumentation was available, scientists gather climate data using proxies like ice cores and tree rings. These proxies can be used to reconstruct ancient temperatures over a variety of temporal scales. In general, the closer to the present, the higher the degree of resolution and precision.

In this chapter, we will look at scientific climate data at time scales closer to the present than most of the examples previously discussed, starting with a relatively coarse-resolution look at Earth's climate for the last 5 million years, then zooming in to a finer-resolution look at climate near the present day. Keep in mind that as the time-scale changes, so do the proxies being used. Also, it is important to remember that, like all scientific data, climate data are inherently filled with "noise," or data points that can sometimes obscure their underlying trend. For example, although winter temperatures are generally at or below freezing in the northeastern United States, warmer days in the 50s, 60s, and 70s (°F) are not unheard of. This is "noise" in winter temperature data, which over short intervals, can make it harder to see the "real" pattern, which is that winters are, overall, cooler than summers.

4.1 The Last Five Million Years

Using fossils from deep-sea sediment cores, which contain climate information in the form of oxygen isotopes (see Box 10), scientists can reconstruct Earth's climate for the last five million years. These analyses show, as in Figure 19, that the climate was warmer than present until approximately 2.8 million years ago. Climate then cooled rapidly and began to fluctuate much more noticeably.

The cause of the change is global cooling associated with the growth of glaciers in the northern hemisphere. The most likely cause of this cooling, however, was far from the high latitudes where glaciers began to expand. This correlates with the time when the Isthmus of Panama formed, creating a land bridge between North and South America, and cutting off oceanic circulation between the Atlantic and Pacific oceans. This strengthened ocean currents like the Gulf Stream, resulting in the transport of more moisture to the northern polar re-

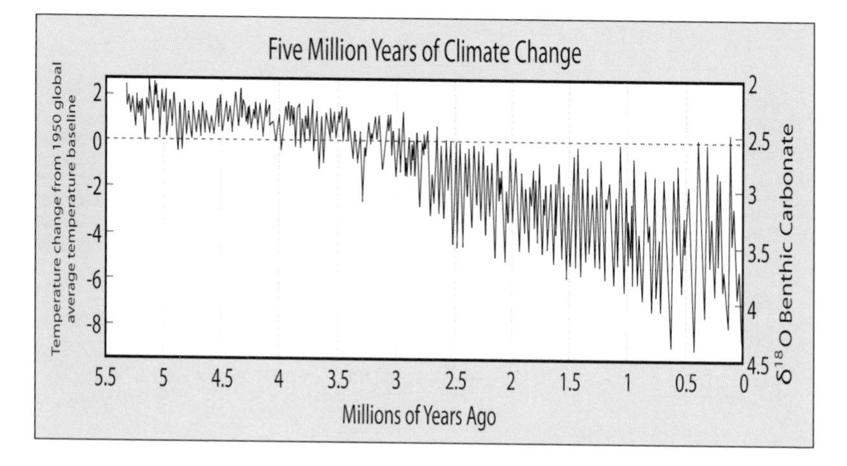

Figure 19. *Five million years of climate change. In the graph, oxygen isotopes from fossil shells of deep-sea marine organisms (Foraminifera) have been used to show relative global temperatures for the last 5.5 million years. The curve is influenced both by the amount of water stored in ice sheets and by temperature at the bottom of the ocean; both cooler temperatures and larger ice sheets cause higher* $\delta^{18}O$ *values. On the left vertical axis are proxy temperature data from an ice core. For more information on temperature anomalies, see Appendix 4. Modified after a graph produced by Robert A. Rohde for Global Warming Art.com.*

gion; this increased snowfall, and was one of the triggers that brought Earth's climate into an **ice age**, causing widespread glaciation and an overall cooling of global climate during the last 2.8 million years.[28]

4.2 The Last 400,000 Years

The orbital variations called Milankovitch Cycles (see Section 2.4.1), which result in changes in incoming solar radiation (insolation) on a roughly 100,000-year cycle, have happened throughout Earth's history, no matter what the overall global temperatures. When Earth was relatively warm, these orbital variations most notably caused changes in precipitation. When the Earth became cooler, however, as happened approximately 2.8 million years ago, the orbital variations resulted in changes in global temperature. Thus, roughly every 40,000 years from 2.8 million to 0.7 million years ago, and every 100,000 years since 0.7 million years ago,[29] ice sheets have expanded into lower latitudes, at their greatest extent reaching the northern parts of what is now the United States.Scientists call these expansions **glacials**, the last one peaking approximately 20,000 years ago in what is called the Last Glacial Maximum. The warmer intervals between glacials, when the ice retreated northward, are called **interglacials**. Earth is currently in an interglacial interval. Prior to the present, the most recent interglacial period occurred approximately 125,000 years ago when temperatures at the poles were 3-5°C warmer than at present, and global sea level was 4-6 meters (13-20 feet) higher than it is today.

Using bubbles and water trapped in ice cores (see Section 3.3), scientists can measure the past atmospheric concentration of CO_2. Atmospheric CO_2 co-varies with temperature in a way that reflects a positive feedback (Figure 20). For example, transitions from a cooling (glaciation) interval to a warming (deglaciation) interval occur due to changes in solar insolation associated with **Milankovitch Cycles**; the warming in turn leads to release of CO_2 from warming ocean water and uncovered soils, increasing atmospheric CO_2 concentrations and further contributing to warming. The reverse occurs in transitions from warm intervals to cool intervals.[30]

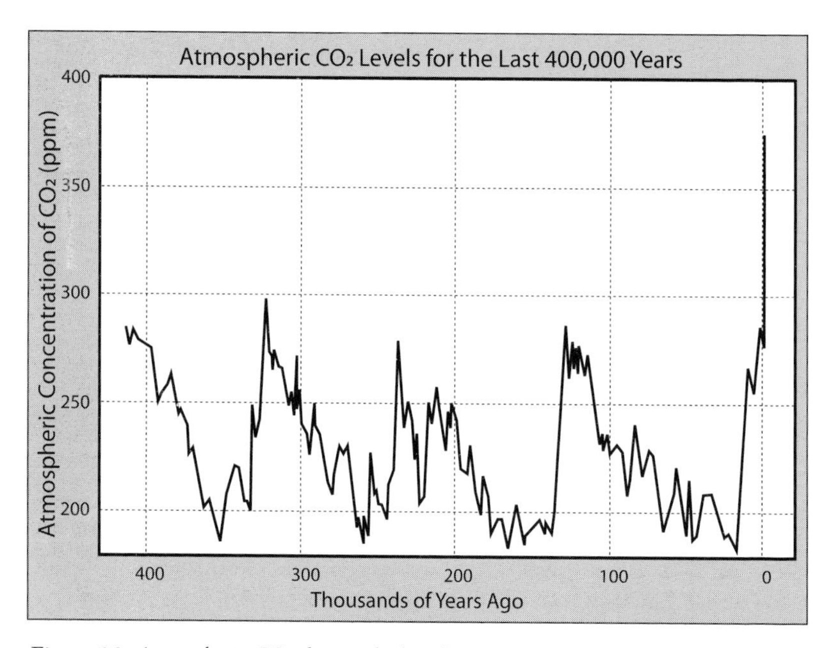

Figure 20. *Atmospheric CO_2 during the last 400,000 years. As was discussed in Box 2, atmospheric CO_2 concentration has been recorded for almost 800,000 years from ice-core air-bubble samples. During that interval, including the last 450,000 years depicted on this graph, the atmospheric concentration of CO_2 has never gone above 350 parts per million until the 1980s, more than 100 years after the start of the Industrial Revolution. Modified after a graph produced by Robert A. Rohde for Global Warming Art.com.*

4.3 The Last 100,000 Years

Using oxygen isotopes from ice cores (see Box 10), scientists can reconstruct Earth's global temperature during the last 100,000-year cycle (Figure 21). Tracing the midpoint of all of the high and low values in the line shows that temperature slowly dropped over this time. Geologically, scientists equate this cooling with the most recent glacial interval, when mastodons and mammoths roamed North America and a thick sheet of ice covered places like New York, Michigan, and northern Europe. This ice began retreating approximately 20,000 years ago, when temperature abruptly increased. If the past is predictor of the future, then without human intervention, global average temperature and atmospheric levels of CO_2 should slowly decrease for the next 60,000 to 80,000 years.

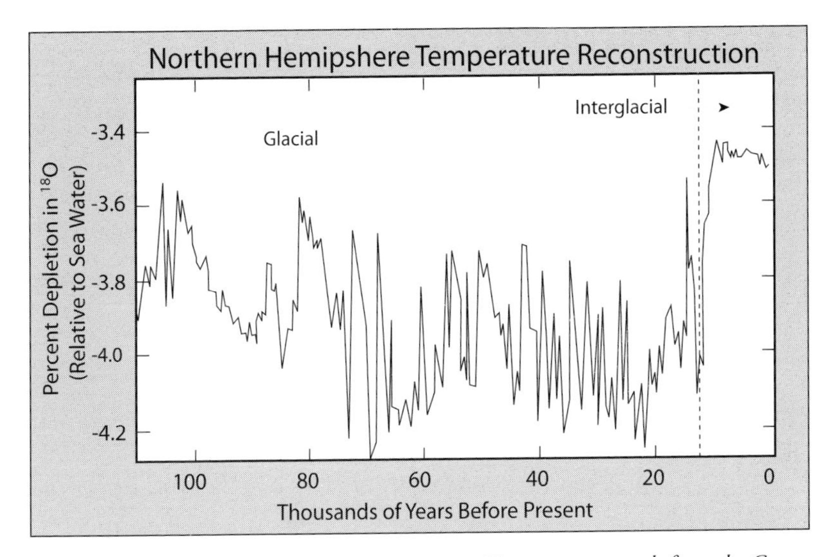

Figure 21. *Greenland ice-core proxy temperature. Temperature records from the Greenland Ice Core based on $\delta^{18}O$ relative to $\delta^{16}O$ for the last 100,000 years. Modified after Broecker, 2003.*[30]

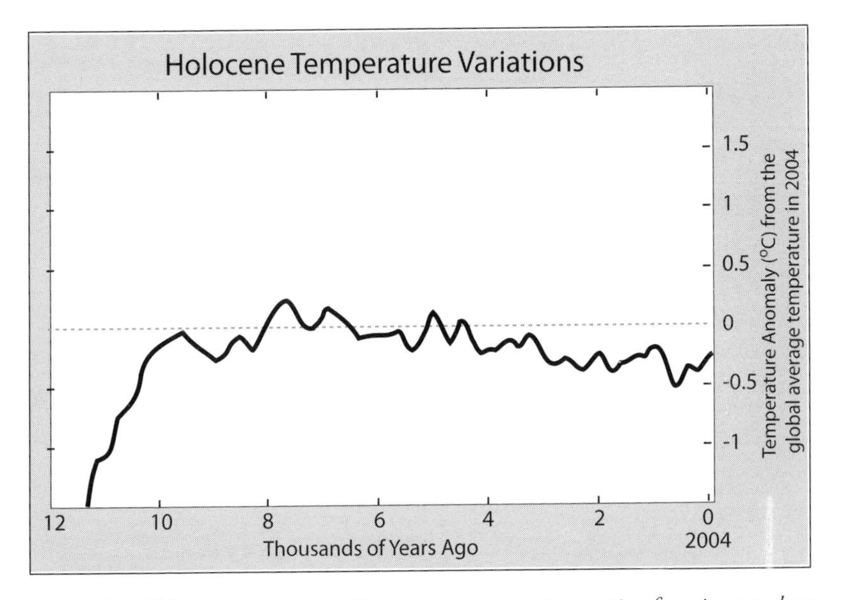

Figure 22. *Holocene temperatures. Proxy temperature reconstruction from ice-core data after the end of the last glacial period, measured in °C difference from the global temperature average in 2004. Modified after a graph produced by Robert A. Rohde for Global Warming Art.com.*

4.4 The Last 10,000 Years

The most recent glacial interval ended approximately 20,000 years ago, and this is visible in the dramatic temperature increase between 20,000 years ago and present (Figure 21). The last 10,000 years or so (the Holocene Epoch; see Appendix 1), however, look dramatically different from the rest of the graph in Figure 21; there appears to be less "noise" in the data.

Figure 22 shows that after the global average temperature increased approximately 10,000 years ago, temperatures in the Northern Hemisphere began a slow, steady decline, which lasted for most of the last 7,500 years. These data (from ice cores from Greenland) support the hypothesis that, at least in the Northern Hemisphere, if natural climate variation were unhindered by humans, temperature would remain nearly constant, or would very slowly decline over the next few thousand years.

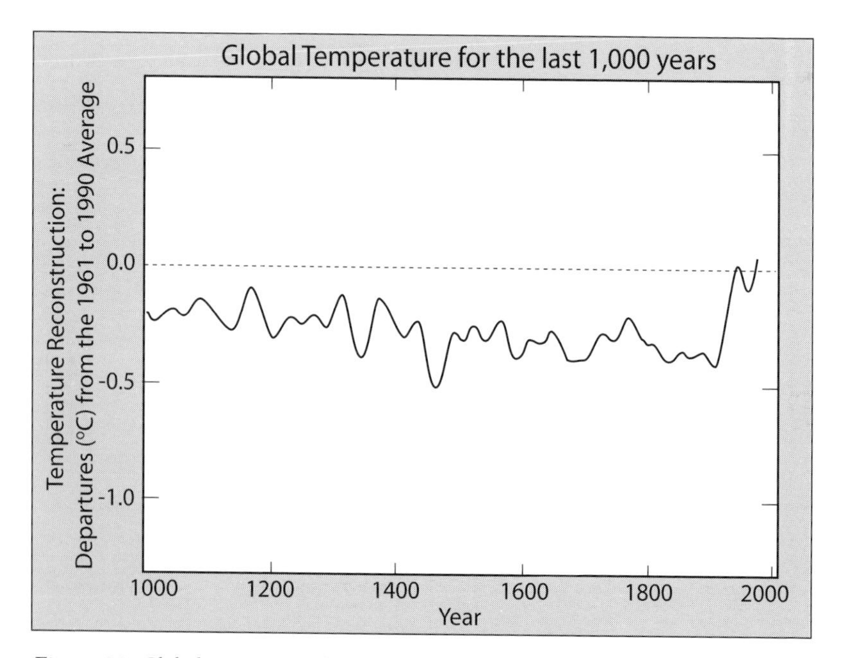

Figure 23. *Global temperature for the last 1,000 years. Global temperature fluctuations in °C for the last 1,000 years relative to the global average temperature from 1961 to 1990, based on oxygen isotopes from ice cores. Modified after Mann* et al., *1998 (see endnote 141).*

4.5 The Last 1000 Years

As scientists look closer to the present, climate resolution continues to improve. As indicated by ice-core temperature proxies, the last 1,000 years show very little significant global average temperature change prior to the 20th century (see Figure 23). A careful eye might notice, however, that the period from year 1000 to 1300 CE is slightly warmer and more "noisy" than from 1300 to present. This represents a time called the **Medieval Warm Period** (see Section 2.4), during which the Viking people inhabited Greenland. In the centuries following (from 1400 to 1800 CE), in contrast, fisherman living in nearby Iceland, where the weather is more temperate, were unable to leave their fishing ports for up to three months out of each year. Although the Medieval Warm Period resulted in only a small increase in global average temperature, it was substantial enough in that region to impact the lives of the people living there.

In the last 100 to 150 years, global average temperature has dramatically increased (Figure 23). This increase is a significant departure from the trend of the last 800 to 900 years, an interval without major anthropogenic influence. In fact, *the warming seen in the last 50 years is*

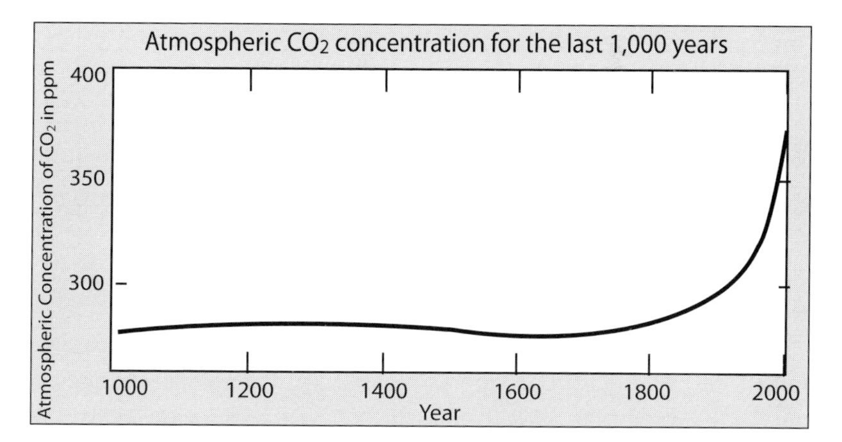

Figure 24. *Atmospheric CO_2 concentrations for the last 1,000 years from data collected from ice cores. Note the rapid rise in CO_2 during the Industrial Revolution (*ca. *1850 to 1900). Modified after a graph produced by Robert A. Rohde for Global Warming Art. com.*

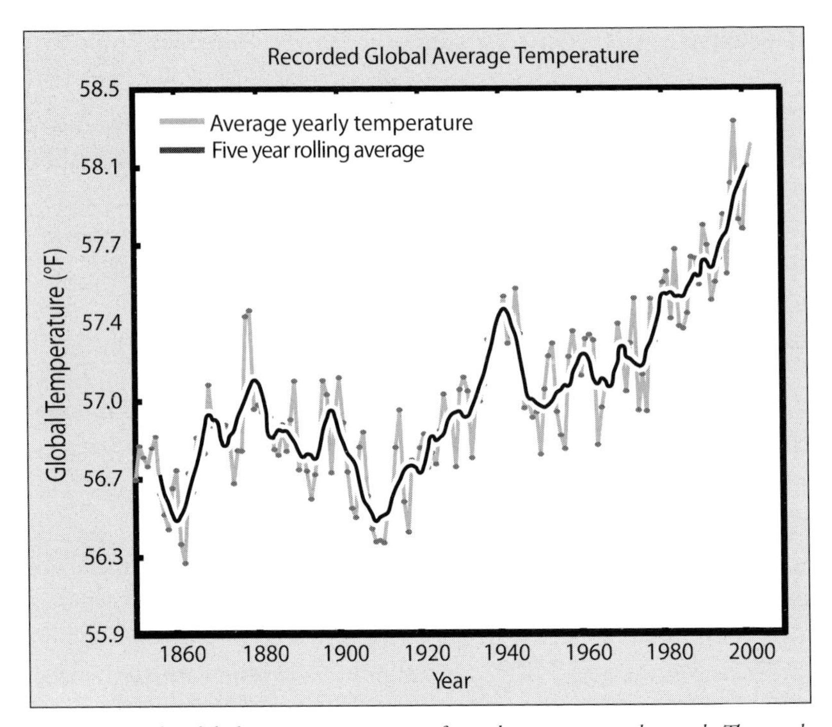

Figure 25. *The global average temperature from the instrumented record. The yearly average shows repeated peaks and furrows, while the five-year average is a much smoother line. The five-year line is a* ***rolling average****, meaning that each year's data point is averaged with those of the two years immediately preceding and two years immediately following. This helps to remove the "noise" so that the overall trend becomes visible. See Appendix 3 for more information on rolling averages. Data from National Aeronautics and Space Administration.*[32]

more pronounced than at any time in at least the last 1,300 years (Figure 24).[33]

4.6 The Last 150 Years

For the last 150 years of Earth's history, scientists have had the advantage of being able to directly measure temperature, without having to rely only on proxy information. These data are called the "instrumental record," because they were recorded by scientific instruments. This record shows the same trend that multiple lines of proxy data have expressed: global temperature is increasing (Figure 25). Because it departs so dramatically from the pattern of the last 1,000 years, this temperature increase is interpreted by most climate scientists as different from what would be expected from natural variation alone.

Even though global average temperature has been increasing over the past 150 years, it has not done so uniformly, every single year, in every part of the globe. Temperature, furthermore, is not the only aspect of climate, and the instrumental record allows scientists, at least in the most recent past, to look at the changes seen in the many other aspects of climate. (This is why the term "climate change" is a more accurate description of what has been occurring than the more familiar term "global warming.") Average global temperature is increasing, causing a reaction in the other aspects of climate, including wind and precipitation patterns. Different regions feel these impacts to varying degrees. For example, North America, South America, northern Europe, and northern and central Asia have already experienced significantly higher precipitation levels. Other areas, like the Mediterranean and southern Africa and Asia, have experienced drier climates than normal.[34]

4.7 The Last 20 Years

The last 20 years of observations have shown even more dramatic changes in many aspects of global and regional climates.

Surface temperature: The great majority of data support the conclusion that the world's average surface temperature has increased over the past 20 years. This is true despite year-to-year fluctuations and regional variations from the global average. It is important to remember that any single year is not indicative of, nor does it disprove, a trend. To understand global surface temperature trends, scientists therefore create graphs with **rolling averages** (see Appendix 3). For example, using a five-year rolling average (Figure 25), the global temperature of the year in question is averaged with the two years immediately preceding and the two years immediately following. So, the global temperature in 2002 is averaged with those in 2000, 2001, 2003, and 2004 to measure the long-term trend, rather than the "noise."

Even without rolling averages, recent temperature data show significant warming. Eleven of the 12 years from 1995 to 2006 rank among the warmest years in the instrumental record (since 1850). Further, during the past century (1906-2005), average temperature has increased by 0.74°C, but warming over the last 50 years (0.13°C per decade) has been nearly twice that for the last 100 years (total temperature increase from 1850-1899 to 2001-2005 is 0.76°C). Widespread changes in extreme temperatures have also been observed over the last 50 years. Cold days, cold nights, and frost have become less frequent, whereas hot days, hot nights, and heat waves have become more frequent.[35]

Ocean temperatures: Observations since 1961 show that the average temperature of the global ocean has increased both at the surface and to depths of at least 3,000 meters (9,250 feet). Such warming causes seawater to expand, contributing to sea-level rise, and also can adversely affect sea life (see Chapter 7).[36]

Snow and ice cover: Global snow and ice cover have declined significantly in both hemispheres over the past two decades, and these decreases have also contributed to sea-level rise. Ice-flow speed has increased for some Greenland and Antarctic glaciers that drain ice from the interior of the ice sheets. Satellite data show that, since 1978, annual average Arctic sea-ice extent has shrunk by 2.7% per decade, with larger decreases during summers of 7.4% per decade. Temperatures at the top of the **permafrost** layer (ground that is frozen year-round) have generally increased since the 1980s in the Arctic (by up to 3°C). The maximum area covered by seasonally frozen ground has also decreased by approximately 7% in the Northern Hemisphere since 1900, with a decrease during spring seasons of up to 15%.[37]

Sea level: Global average sea level rose at an average rate of 1.8 millimeters per year between 1961 and 2003. The rate was even faster (3.1 millimeters per year) between 1993 and 2003, but whether this reflects an increase in the longer term trend is unclear. The IPCC concluded in their 2007 report that there is "high confidence that the rate of observed sea level rise increased from the 19th to the 20th century."[38]

Precipitation and drought: There is considerable evidence of global changes in evaporation and precipitation, probably linked with higher temperatures, which have contributed to more intense and longer droughts over wider areas since the 1970s, particularly in the tropics and subtropics. These changes are the result of **freshening** (a decrease in salt content in ocean water) of mid- and high-latitude ocean waters because of increased precipitation, together with increased salinity in low-latitude waters caused by decreased rainfall. Changes in sea-surface temperatures, as well as wind patterns and decreased snow cover, have also been linked to droughts. The frequency of heavy precipitation events has also increased over many land areas.[39]

Summary: Climates of the Recent Past

Every major aspect of climate that is measurable has shown significant change since the beginning of the Industrial Revolution, and these changes are of a magnitude and rate not seen for more than at least 1,300 years. There is no longer any question among the majority of qualified scientists that climate change is occurring. As the 2007 IPCC report put it: "Warming of the climate system is unequivocal."[40]

Figure 26. *Stora Enso Pulp and Paper Mill in Nuottasaari, Oulu, Finland. Stora Enso was one of the first companies in the forest products industry to calculate the carbon footprint of its operations, and is striving to reduce its* CO_2 *emissions worldwide by 20% by the year 2020.*[41] *Photograph by Estormiz via Wikimedia Commons.*

5. HUMANS AS A CAUSE OF CLIMATE CHANGE

The idea that humans might be responsible for climate change was first proposed in the 1890s by a Nobel laureate chemist, Svante Arrhenius, who pointed out that continued use of coal, oil, and other fossil fuels by humans could change the atmosphere's composition and warm the Earth.[42] The idea started being taken seriously in modern times in the late 1970s, when a small number of climate scientists began to suggest that human activities, especially those resulting in emissions of CO_2 and other greenhouse gases, were actually changing. As more research was done, this idea began to gain more support. By the early 1990s, evidence from an increasing variety of climate data were supporting previous theoretical predictions. The 1995 IPCC report stated rather cautiously that "the balance of evidence suggests that there is a discernable human influence on global climate."[43] The 2001 IPCC report took a step further, declaring that "notwithstanding some role for natural variability, humans are almost certainly warming the planet right now."[44] The 2007 IPCC report more strongly concluded with "very high confidence" that "the global average net effect of human activities since 1750 has been one of warming."[45]

What is the basis for these increasingly confident conclusions? The evidence can be grouped under five headings (see also Box 2):

(1) CO_2 is a greenhouse gas.

(2) CO_2 has been increasing in the atmosphere since the beginning of the Industrial Revolution (*ca.* 1850).

(3) Temperatures have risen along with CO_2 levels.
(4) Climate models cannot be made to accurately represent past and present climates without human forcing.
(5) No other known cause appears adequate.

If, as the IPCC and many scientists contend, the evidence for each of these conclusions is overwhelming, why does controversy over the cause of climate change persist? Massachusetts Institute of Technology climatologist Kerry Emanuel suggested this answer:

> *If one could change the concentration of a single greenhouse gas while holding the rest of the system (except its temperature) fixed, it would be simple to calculate the corresponding change in surface temperature. ... Almost all the controversy arises from the fact that in reality, changing any single greenhouse gas will indirectly cause other components of the system to change as well, thus yielding other changes...*[46]

In other words, because climate is so complex, it is impossible to do a typical style of experiment, in which only one variable is allowed to change. So it is impossible to get the kind of clear demonstration of causality that many people expect of science.

5.1 CO_2 is a Greenhouse Gas

As discussed in Chapter 2, CO_2 is a greenhouse gas. It is of particular importance because it is a gas currently being emitted in large amounts by human activity, and once released into the atmosphere, it remains there for hundreds of years. Carbon dioxide is emitted by many natural processes, including volcanism, and has many natural places of storage, including the oceans. Increasing anthropogenic CO_2 emissions, however, due to the burning of fossil fuels, changes what is usually naturally balanced.[47] To help understand this intricate balance, scientists use the concept of **radiative forcing**. Radiative forcings are factors that affect the amount of energy received by or lost from the Earth, such as atmospheric gases, clouds, and albedo.

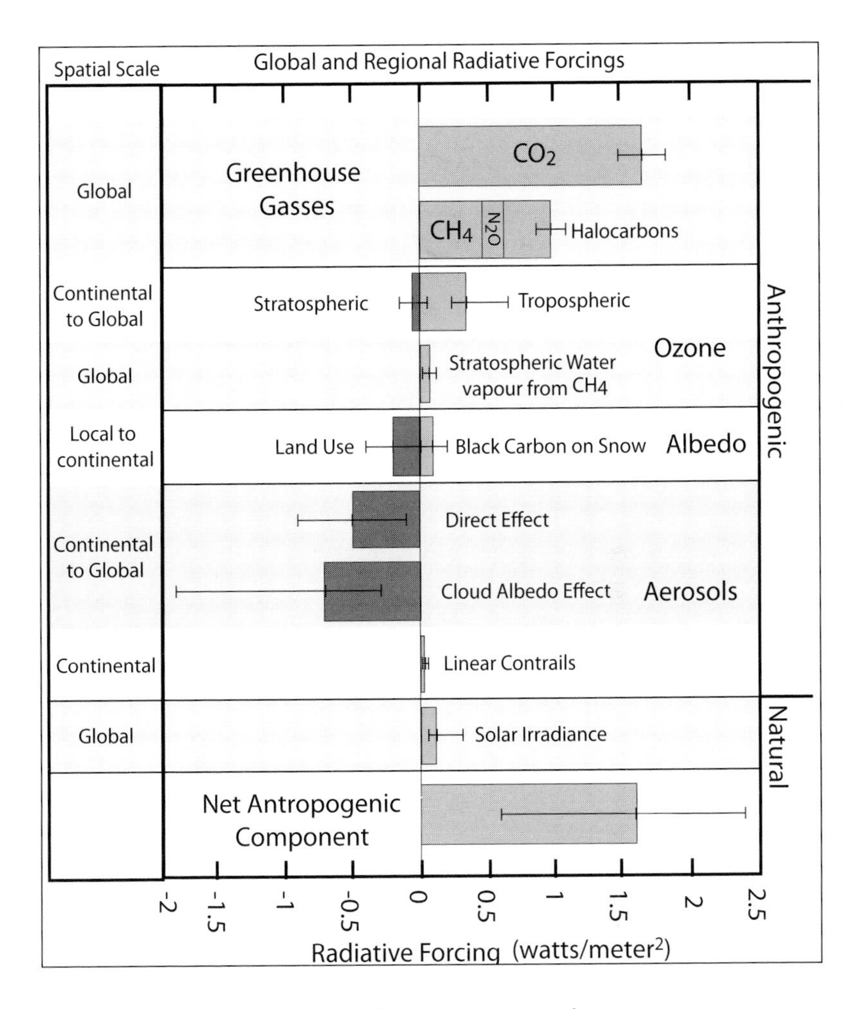

Figure 27. *Radiative forcings. The Sun's energy, measured in watts per square meter, can be reflected back into the solar system or absorbed by the Earth and its atmosphere in the form of heat energy, and the absorbed energy can be radiated back in space in the same way that your body loses heat on a winter day. Light gray bars to the right of "zero" indicate positive (warming) forcing, whereas dark gray bars to the left show negative (cooling) forcing components. When the physical forcings reflect or otherwise lose their energy to space, it is a negative forcing, When the forcings absorb the energy, they are positive. In the diagram, the majority of the forcings are anthropogenic, but some are natural. Overall, with anthropogenic forcings added, there is a positive radiative forcing, which results in a general temperature increase. Modified after a graph from IPCC, 2001 (see endnote 44).*

When incoming and outgoing radiation are in balance, a condition called neutral radiative forcing exists; this is represented by the "zero" line in Figure 27. Most of the forcing components shown in Figure 27 are anthropogenic (human-caused). These include emissions of CO_2, methane (CH_4), and nitrous oxide (N_2O), among others. Compared to the positive forcing from the Sun (insolation), anthropogenic positive forcings are much higher, and the forcings caused by human activities push radiative forcings in the positive direction, meaning greater warmth (for more on forcings, see Box 6).

5.2 Anthropogenic Increases in Greenhouse Gases

In 1958, Charles Keeling of the Scripps Institute of Oceanography began a project repeatedly measuring the CO_2 content of the atmosphere on top of Mauna Loa in Hawaii. Mauna Loa was chosen because the air blowing over its summit – from the western Pacific – is relatively free of local sources of pollution. The result of these measurements is called the **Keeling curve** (Figure 28).

As summarized in Box 4, atmospheric levels of CO_2 prior to 1958 come primarily from the analysis of air bubbles in ice cores. Together with measurements of modern atmospheric conditions, the data indicate that the global atmospheric concentration of CO_2 has increased from a pre-industrial value of approximately 280 parts per million (ppm) to approximately 379 ppm today. *This change greatly exceeds the range over at least the last 800,000 years* (180-300 ppm) as determined from existing ice cores (Figure 24). The annual CO_2 concentration growth rate was larger during the period 1995-2005 than it has been since the beginning of continuous direct atmospheric measurements in the 1950s.[48]

Carbon dioxide is notthe only greenhouse gas increasing in concentration. According to the 2007 IPCC report,[49] the increase since 1850 in CO2, methane, and nitrous oxide together provide greater radiative forcing than at any other time in the past 10,000 years (and thus probably well back into the Pleistocene).

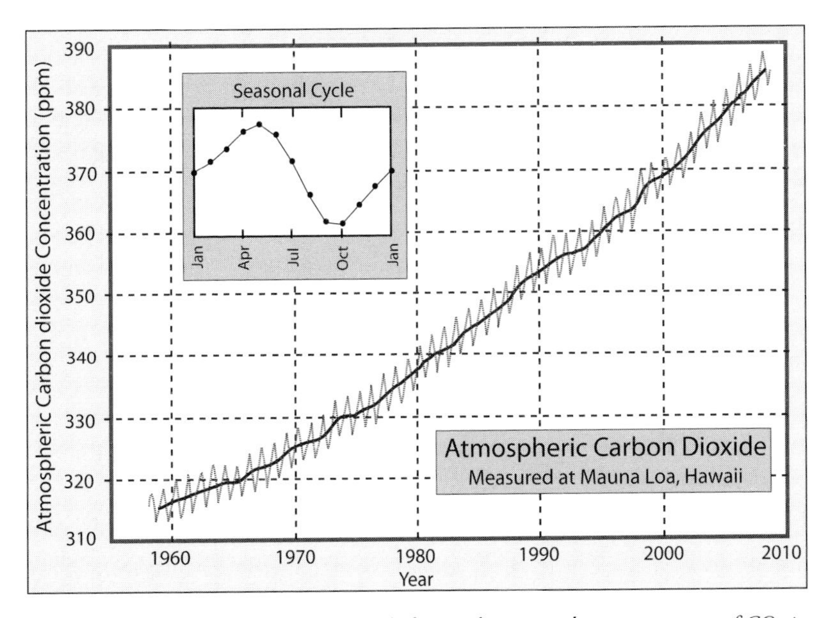

Figure 28. *The Keeling curve. This graph depicts the rise in the concentration of CO_2 in the atmosphere as recorded on top of Mauna Loa, Hawaii, since before 1960. The small annual oscillations seen in the graph are enlarged in the box on the left side of the graph and represent the slight decrease in CO_2 annually due to the leafing and subsequent loss of leaves in deciduous trees in the Northern Hemisphere that (because of the concentration of forested landmasses north of the equator) is much stronger than the opposite effect in the Southern Hemisphere. Modified after a graph produced by Robert A. Rohde for Global Warming Art.com.*

Burning fossil fuels: The largest source of anthropogenic CO_2 is the burning of fossil fuels – oil, coal, and natural gas. All of these fuels are taken from natural carbon sinks that have trapped the carbon for millions of years. Fossil fuels are burned for transportation (cars, trucks, ships, trains, planes), to produce electricity, and to heat homes and other buildings. Annual CO_2 emissions from these sources increased from an average of 6.4 billion tons of carbon in the 1990s to an average of 7.2 billion tons of carbon during 2000 to 2005.[50]

Cement making: Cement is a fine powder used to make concrete, a widely-used building material made primarily of limestone ($CaCO_3$), which, like fossil fuels, is a natural carbon sink. Cement is made by grinding limestone, mixing it with additives like sand and clay, and heating it all to an extremely high temperature to cause a chemical reaction called **calcination**. Carbon dioxide is released during the calcination process (and also from combustion of fuels to heat the kiln). The cement industry contributes approximately 5% to global anthropogenic CO_2 emissions.

Agriculture is primarily responsible for anthropogenic increases in two other greenhouse gases – methane (CH_4) and nitrous oxide (NO_2) (see Table 2). Agriculture releases methane primarily from rice cultivation and the fermentation that occurs in the digestive tracts of cows. Nitrous oxide is released mainly as a side effect of application of fertilizer. Global atmospheric concentration of methane has increased from a pre-industrial level of approximately 715 parts per billion (ppb) to approximately 1,774 ppb in 2005. These levels of methane in the atmosphere are much higher than the natural range (320-790 ppb) during the last 50,000 years as documented from ice cores. Agricultural practices also have increased the global levels of the atmospheric concentration of nitrous oxide, from 270 ppb in pre-industrial times to approximately 319 ppb presently.[51]

Deforestation: Forests are carbon sinks, absorbing and storing CO_2 and producing oxygen.[52] When they are removed for timber, cattle grazing, or other agricultural uses, or burned in a fire, the carbon that was stored in the forests is released directly into the atmosphere. Of course when trees fall, die, and decay naturally, they also release their stored carbon, but human-caused deforestation happens much more quickly and at a vastly greater scale than natural processes.[53]

Human-caused destruction of both temperate and tropical forests also negatively impacts the environment in many other ways. After deforestation, soil is lost through weathering, along with the nutrients in the soil. Grazing activities by cattle compacts soil and lowers the ability of the soil to capture and retain water. Each of these also have indirect affects on the carbon cycle, precipitation, and other climate variables.

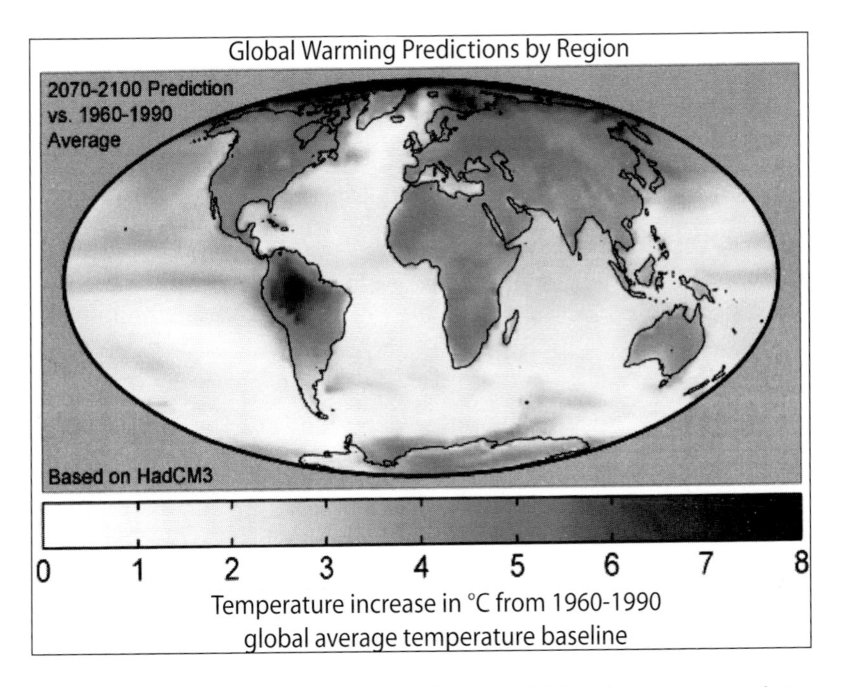

Figure 29. *Regional predicted temperature change model. Based on current population growth and greenhouse gas emissions, this map shows model predictions of which areas of the globe will feel the greatest degree of warming during the last thirty years of the 21st century. The temperature anomalies are relative to the global temperature average of 1960-1990. Areas that are darker in color will be more greatly impacted by temperature changes than areas that are lighter. Some light areas might even feel slight cooling. Modified after a graph produced by Robert A. Rohde for Global Warming Art.com.*

5.3 Models

The observed patterns of warming, including greater warming over land than over the ocean, and their changes over time, are simulated successfully only by models that include anthropogenic forcing. The ability of coupled climate models to simulate the observed temperature evolution on each of the six continents, then averaged together to form a global land and global sea change, provides strong evidence of human influence on climate (Figure 29). Climate modeling is discussed in more detail in Chapter 6.

5.4 Alternative causes

Anthropogenic increases in CO_2 are not the only possible cause of current climate change. The primary alternative is changes in solar radiation, or energy from the Sun. The total amount of solar energy reaching Earth can vary due to changes in the Sun's output, such as those associated with sunspots, or to changes in Earth's orbit (see Chapter 2). These variations occur on timescales from millions of years through to the more familiar 11-year sunspot cycles.

Scientists who study the Sun believe that our star emitted one-third less energy approximately 4 billion years ago than it does today, and that it has been steadily brightening ever since. Yet for most of this time, Earth has been even warmer than today, a phenomenon sometimes called the **Faint Young Sun Paradox** (see Box 9). The reason: in the geologic past, higher levels of greenhouse gases trapped more of the Sun's heat, raising and maintaining Earth's temperature.[54]

Based on studies that had been done since its previous report, the IPCC in 2007 halved its earlier estimate of solar forcing on global temperature increase over the past 250 years, from 40% to 20%. Even if this understates solar forcing over this time interval, there is no evidence for correlation between solar activity and warming over the past 40 years (direct measurements of solar output since 1978 show rises and falls coincident with the well-known 11-year sunspot cycle, but no overall trend up or down). Similarly, as already mentioned (see Section 2.4.1), there is no trend in direct measurements of the Sun's ultraviolet output. Thus, for the period for which we have direct measurements, Earth has warmed significantly even though there has been no corresponding rise in solar activity.

Summary: Humans as a Cause of Climate Change

Anthropogenic greenhouse gas emissions are produced through the burning of fossil fuels for energy consumption, cement making, agricultural practices, and land use changes like deforestation. These activities have a positive radiative forcing and can contribute to warming of the Earth. There are other, natural processes that have contributed to global climate change in the past, but positive radiative forcings from those activities were usually balanced out by naturally occurring negative radiative forcings (see Section 5.1 and Figure 27).

It is difficult to correlate and predict the changes in future climate in specific times and places based on radiative forcings because of the complexity of the climate system. Locally, there are many uncertainties and small variations in forcings that affect regions differently. Global averages also mask changes like extremely hot nights or extremely cold nights, which have likely increased in the past decade because of anthropogenic forcings. So although scientists understand that all aspects of the climate system are interconnected, the intricacies of their relationships with additional radiative forcing is the topic of much research. One thing is certain – additional radiative forcings increase the temperature of the Earth, and this impacts every other aspect of the climate system in ways that we are still discovering. Anthropogenic climate change is a scientific experiment on our planet with an unknown outcome.

Figure 30. *New Orleans, Louisiana, 31 August 2005. People walk through flood waters to get to higher ground in the aftermath of Hurricane Katrina. Photograph by Marty Bahamonde, U.S. Federal Emergency Management Agency.*

6. CLIMATES OF THE FUTURE

6.1 The Problem of Prediction

What is a **prediction**? In common terms, it means an educated guess about what might happen in the future. Anyone can make such a guess based on a variety of factors, from feelings and hopes to analysis of objective information. Scientific predictions, however, are based only on observational data and logical analyses, and are usually much more specific than nonscientific predictions. But how exactly does scientific prediction work? For example, when the weather forecaster predicts that tomorrow will be sunny with a 30% chance of rain, where does such a statement come from?

Let's look at the example of weather forecasting in a little more detail. We all know from practical experience that weather forecasters cannot predict weather more than a few weeks in advance. This is because complex systems like the weather are influenced by a huge number of small cause-and-effect relationships. Because a change in one aspect of the system impacts the others, the fewer potential changes that there are, the greater the certainty will be. This means that the farther ahead we look, the larger the number of natural variables that influence the weather and the more opportunity there is for variables to vary. As we look at the nearer-term future, for example at weather a day or two in advance, there will be greater certainty in the prediction, because the number of variables is smaller, and more of those that remain become fixed.

Climate prediction suffers many of these shortfalls as well. Because of this, scientists use **models** to predict both weather and climate. Models are formal statements of how we think a process or phenomenon happens, usually expressed in quantitative terms. We all use models every day to understand, and predict, events around us. For example, when we are waiting to walk across a street and see a car approaching, we want to predict whether it will be safe to cross before the car reaches us. To do so, we observe the speed of the car and then mentally calculate how long it will take to reach us at that speed. We compare that time interval to our estimate of how long it will take us to cross. If we think that it will take us longer to cross than it will take the car to reach us, we then calculate whether, if the driver of the car sees us and decides to slow down or stop to let us cross, the car can actually slow or stop in time. We also, if we can, make observations about the driver to help us decide whether he or she will notice us.

All of these judgments – which are made in a second or two in our heads without our really thinking about it – are exactly the same kinds of steps that a scientist uses to construct a model of a system that he or she wishes to understand and predict. All models are simplifications, that is, they exclude some of the variables in the system that they are evaluating, *e.g.,* the model of the car and kind of tires on the car, which might influence the events in some subtle way. And they are built knowing that not all relevant information will be available (*e.g.,* do the car's brakes work, or is the driver fatigued?). As a scientist takes more and more variables and information into account and includes them in the model, the model becomes more and more complex. Our model might become more accurate, but it will also be more difficult to interpret.

This does not mean that complex processes or phenomena are impossible to understand using complex models. It does, however, mean that the purpose of the model needs to be carefully considered. In our street-crossing example, an acceptable result of our mental modeling might simply be that we get across the street without being hit by the car, even if this means we have to run when we get to the middle. For other models, we might want or need much more precise predictions. If, for example, we want to predict what the temperature will be at a certain point in the future, we might use the more complex model even if it sacrifices simplicity.

To attempt to accurately model weather or climate, scientists must evaluate relationships among a huge number of variables, including the flow of oceans, atmospheric wind patterns, cloud formation and cloud reflectivity, heat transfer through the atmosphere and oceans, and heat absorption and reflectivity by various surfaces on the Earth (ice, water, rock, sand, vegetation). For this reason, modeling of global climate using computers has been called "perhaps the most complex endeavor ever undertaken by mankind," requiring millions of lines of computer instructions written by teams of scientists over the course of years.[55]

6.2 How Weather and Climate Models Work

Let's return to the problem of weather prediction. Imagine that someone is wondering what the weather will be later today to plan an outing. That person might look into the sky and notice whether any clouds seem to be moving in their direction. She might interpret a sudden wind accompanied by gathering clouds moving across the sky as a signal of coming rain. By doing so, she is making a simple weather model. It is a model because it is a simplification of reality based on a limited number of cause-and-effect relationships; she might know that the wind patterns are complex, but she generalizes that the clouds are moving in her direction. She also knows from experience that she cannot be 100% sure that it will rain, and although she might not consciously assign a probability to it, she might decide that there is a large enough chance of rain that she decides to make alternative plans.

Weather and climate models do the same thing as the everyday example above, but they do it mathematically. Using numbers and equations works better than rules of thumb not just because mathematics are more precise, but because scientists can communicate very specifically how they came up with their estimates; they can use known physical laws and statistical relationships, and compare with great specificity which models work best and try to figure out why.

At the most basic level, models of both weather and climate work by estimating quantitatively how air moves from one place to another and how other processes interact with it as it moves. Let's say, for example, that someone is creating a weather model for the northeastern U.S. She needs to have some simple boundary around the area that she is modeling, so she imagines a big cube of air over the geographic area in which she will make her predictions (Figure 31). She wants to know how air is moving through this cube (*i.e.,* across the area). The easiest way to do this is to divide the big cube into smaller cubes: in two dimensions, these are squares like a checkerboard across the landscape, and in three dimensions, these are vertical stacks of cubes in the atmosphere like the layers of a cake.

Imagine now that each small cube has its own weather station that captures the average weather conditions within that cube. Each station will keep track of weather conditions in its cube, to track how those conditions might influence the weather conditions in surrounding cubes. The model will start with certain weather conditions (which modelers call "initial conditions"), for example, what the weather is right now (Figure 31). Variations in air conditions among the cubes and "forcings" from outside the large cube (*e.g.,* heating from the Sun) will cause air to move. The modeler can estimate how fast the air moves from cube to cube horizontally across the landscape. She assumes by physical laws that air cannot pile up in one cube, and disappear from another; therefore, if some air moves into one cube, it must move out at the same rate into other cubes. A set of standard physics equations allows her to estimate how the air moves around, from small cube to small cube, all within the big cube. The size of the cubes (that is, the spatial and vertical resolution) controls how detailed the model results will be.

Suppose now that the modeler is interested in predicting the weather conditions at each of his virtual weather stations at some interval of time later, perhaps the weather for tomorrow. Without a model, she could calculate how far an air mass (for instance, a snowstorm or a hurricane) will move over the course of a day, knowing its speed

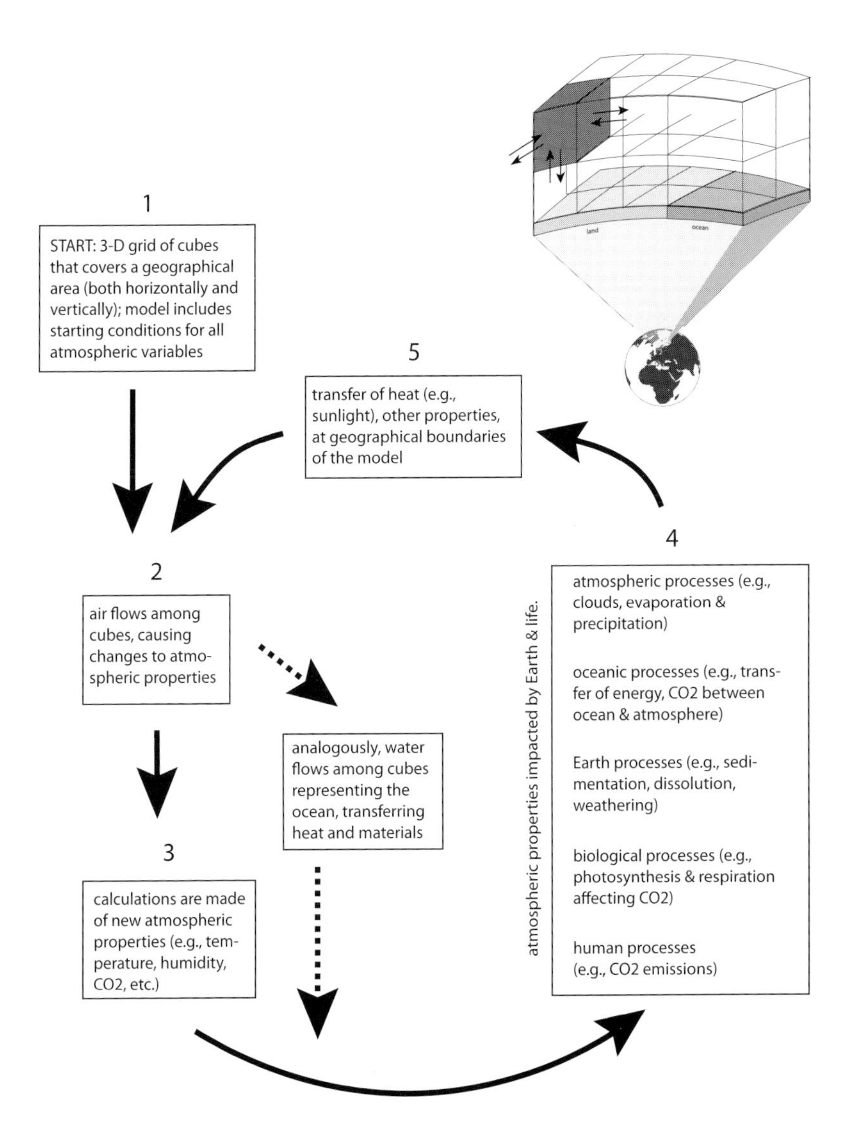

Figure 31. *Making a climate model. This flowchart is a generalization of the steps involved in a numerical weather or climate model; individual models vary depending on the geographic area and time scale being modeled and of course the scientific questions being asked. The general idea is that mathematical equations estimate movement of air and water, their physical properties, and changes in their chemistry. The loop in the diagram includes all of the calculations made during a "time step," which could be from minutes to many years. Global climate models take into account both atmospheric and oceanic circulation, because these have many important interactions. Models that represent long intervals of time might include geological process such as weathering, whereas other models might ignore long, slow processes, but estimate short-term processes such as diurnal (day-night) and annual cycles.*

and direction. But for several reasons (one of which is that weather processes are nonlinear, meaning that simple changes produce complex results; note the often complicated tracks of real hurricanes), it is more effective to recalculate repeatedly at small time intervals, *e.g.*, every ten minutes.

She also needs to remember, of course, that there is no such thing as an isolated atmospheric cube. The "boundary conditions" (such as topographic features and temperatures along the boundary of the cube) during the model simulation could come from the data of a model that covers a larger area, perhaps even a global model. Larger scale models have coarser spatial and temporal resolutions, and the boundary condition estimations get less accurate as the model is run for longer intervals of time.

A model of the northeastern U. S. might have cube sizes that are too large (at least 10 kilometers by 10 kilometers horizontally, by several hundred meters vertically) to capture some processes that occur at smaller scales. An example might be air flow associated with a developing thunderstorm. The model can take these limitations into account by estimating the average affects of the sort of conditions that lead to thunderstorms, even if it does not simulate the actual air movements in detail. That is, the model could estimate rainfall and temperature change in a particular cube using other calculations (called **parameterization**) based on existing climate conditions in each cube. An imperfect analogy might be using a large-scale map to find a location; once your search is at higher resolution than the map can provide, you look around and use other kinds of information to find your exact destination.

Global climate models are much like the smaller-scale model discussed above. Because they cover the entire globe, the geographical area, or "grid size" (such as the system of cubes described above; Figure 31), must be larger. For example, the newest version of a widely-used global climate model[56] is a grid of 1,070 cubes from the North Pole to the South Pole and 1,440 cubes around. That means that each cube of information is approximately 25 by 30 kilometers in size. It has 24 layers in the atmosphere and 50 layers in the ocean, and each of those have thinner layers closer to the surface because of the detail of the topography and processes closer to the surface. Taking all three directions into consideration, these models have much higher spatial resolution than the models of the 1980s that provided some of the first important information on climate change. In principle, the smaller the grid size and the smaller the time interval, the more accurate the model can be; the resolution is ultimately limited by the speed of the world's fastest computers.

The number of calculations is large not just because the atmosphere is so complex, but also because models must include more than just the atmosphere. The ocean, for example, is an important part of the climate system, therefore global climate models must also take ocean circulation into account, which is also very complex. The ocean carries a considerable amount of heat from the equatorial regions toward the poles and is the source for the water in much of the world's precipitation. It is likely that changes in ocean circulation, both in surface currents and in **up-** and **downwelling** of ocean water, has itself forced climate changes over much of Earth's history.[57]

Just as important, global climate models take atmospheric chemistry into account, including of course the role of the carbon cycle in controlling atmospheric CO_2 concentrations (see Box 4). The ocean both takes up and gives off CO_2 in different amounts according to local oceanic and ecological circumstances. The carbon cycle must be represented in climate models by ecological models that estimate the quantity of carbon used in photosynthesis on land and in water bodies, and the ways in which the carbon in living organisms (organic carbon) is recycled back into atmospheric CO_2. The limit on the number of natural processes that can be incorporated into the models is limited by some combination of our scientific knowledge of those processes and by available computing power.

6.3 Model Results

One way to test whether a climate model is useful for prediction is to explain a pattern from the past, like observed temperature changes in the past century. Figure 32 shows the results of models that generate predicted temperature changes with and without the increase of

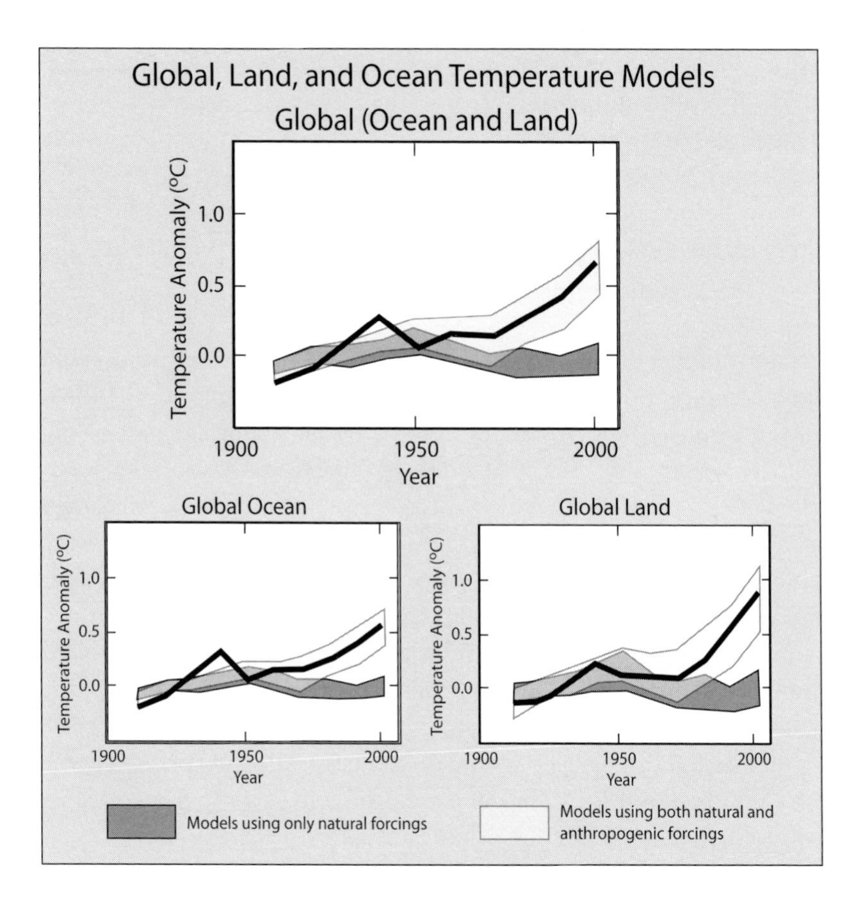

Figure 32. *Global, land, and sea climate-model predictions. Models can also be used to contrast different assumptions about how the Earth system works, comparing different model scenarios with existing data. In these plots, both global land and sea temperature averages over the last 100 years have been compared with two model scenarios, one with anthropogenic forcings such as* CO_2 *emissions (light gray bands) and one with only natural forcings (dark gray bands). The actual data (black lines) match the model with anthropogic forcings much better, giving scientists added confidence that these forcings are responsible for increased temperatures over that interval. Modified after IPCC, 2007 (see endnote 7).*

CO_2 due to human activities. The model results are shown as bands, because the data represents five different models run under a variety of assumptions. Decadal averages of real data are shown as a black line. There are three plots (land, ocean, and combined) because temperatures of ocean and land react somewhat differently to changes in cli-

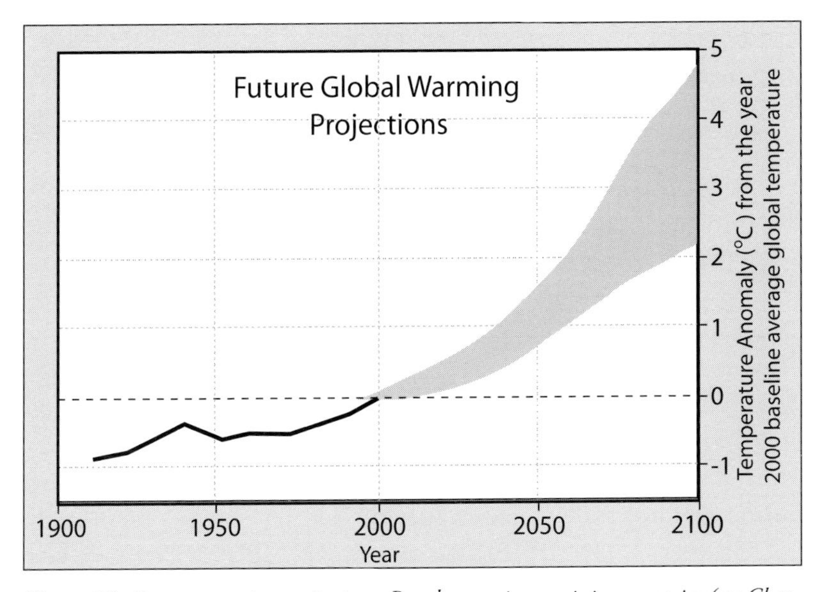

Figure 33. *Future warming projections. Based on various emission scenarios (see Chapters 6 and 7), it is predicted that global average temperature could rise anywhere from just over 2°C to approximately 5°C relative to a baseline temperature from 2000. Modified after a graph produced by Robert A. Rohde for Global Warming Art.com.*

mate. One thing to note from these plots is that the data on observed temperatures is quite similar to modeled temperature change using "anthropogenic forcing" (additional CO_2 from fossil-fuel burning), giving improved confidence that anthropogenically released CO_2 is a major cause of the observed temperature changes to date. All of the natural causes of climate change combined would likely have produced a global temperature increase of no more than 0.1°C, compared to an observed change of greater than 0.5°C.

Another way to use climate models is to make predictions about the future. These predictions are often for changes in global temperature over the next 100 years (to the year 2100), and the results can be depicted as average global temperature through time or as temperature distributions geographically (Figures 29 and 33). The average, or zero-level, temperature in models presented here is the average global temperature from 1960 to 1990. Common to all results is that future temperature changes will not be uniform over the Earth's surface, and that temperature increases will be greater on land than in the ocean. This makes sense, because water heats and cools much more slowly than either air or land.

Figure 33 shows the spectrum of variation among eight models from different climate research centers around the world; all assume a world with sustained economic growth and no significant action to combat climate change, but they vary in details of growth and energy use. This world in 2100 would have 15 billion people who are not very efficient about energy use, and who use moderate levels of fossil fuel (mostly coal). The predictions are relative to temperatures in 2000. The estimates of change using these various models yield approximately 2° to over 4°C. Such changes might not seem like a lot, but on a global scale would change the Earth's climate significantly (see Chapter 7). It is striking that models with different strengths and weaknesses, created by different groups of researchers with different ideas about how the Earth system works, nevertheless give consistent results.[58]

All of the climate models in Figure 33 predict rather similar temperature estimates through approximately the year 2030 – that warming will be approximately twice as large as that which occurred during the 20th century. The models diverge, however, as they estimate out to the year 2100. The range of models suggests that the globe will continue to warm from a minimum change of 1.8°C (the most likely range is 1.1° to 2.9°C), to a high scenario of approximately 4.0°C (the likely range is 2.4° to 6.4°C). Even the low range is a larger increase than was actually observed during the 20th century.

Models also show that warming is expected to vary geographically (Figure 29). The most significant changes are predicted to be in northern South America and near the North Pole, where temperatures could be as much as 8°C higher in 2100 than they are at present. Contraction of the Greenland Ice Sheet is projected to continue to contribute to sea-level rise after 2100, with losses through melting with increased temperature occurring more rapidly than gains due to increased snow precipitation. In contrast, the global models estimate that the Antarctic Ice Sheet will remain too cold for surface melting and (perhaps counterintuitively) will more likely gain in mass due to increased snowfall.

Note that, based on these models, even if greenhouse gas and aerosol concentrations in the atmosphere had remained constant from the year 2000, we would still see approximately half of the expected temperature change. In effect, the Earth is still warming from greenhouse gases added during the 20th and late 19th centuries. This is an extremely important point: *it will take at least 100 years to stabilize global temperatures after we stop increasing the atmospheric concentration of* CO_2, and several centuries to stop the thermal expansion of the ocean, which raises sea level by another meter above that caused by glacial melting.[59]

In this sense, at least some global climate change is inevitable and cannot be "stopped." It can, however, still be slowed and perhaps even reversed, at least in part, but the longer we wait to take action, the more difficult this will become. Many climate scientists fear that there might be a point beyond which climate change cannot be reined in, a threshold past which effectively permanent changes (at least on human time scales) will be unavoidable. Because of the complexity of the climate system, however, there are so far no reliable estimates of where such a "tipping point" might be.[60]

Summary: Climates of the Future

Some of the world's most complex computer models are being used to predict the impact of increasing CO_2 in the atmosphere over the next century. Scientists have confidence that these models are reasonably effective, mainly because they can be used to reproduce climate variations that have actually been observed over the past century.

Although a number of teams of scientists have independently created such models, and the models vary in their strengths and weaknesses, all predict the same thing – that global temperature will rise several degrees Celsius by 2100. These climate models also show that climate change is not geographically uniform: warming will be much more dramatic in some areas than others, and climate change in some areas could more significantly involve changes in precipitation than changes in temperature. Climate models have become more sophisticated over the past 30 years as both the scientific understanding of climate and computing power have grown, yet the essential conclusions about the role of CO_2 as a greenhouse gas have remained fairly constant.

Many climate scientists believe that some amount of climate change is already inevitable, that even if anthropogenic greenhouse gas emissions were to decline dramatically and immediately, global temperatures would still rise by at least 1-2°C.

Figure 34. *Artist's concept of NASA's planned Orbiting Carbon Observatory. The mission would have included the first spacecraft dedicated to studying climate change. It was designed to provide the first global picture of the human and natural sources of* CO_2 *and the places where this important greenhouse gas is stored, which in turn would improve global carbon cycle models, forecasts of atmospheric* CO_2 *levels, and predictions of future climate change. Unfortunately, the launch in February 2009 failed to reach orbit, and the satellite was destroyed. Future plans are uncertain. Graphic by National Aeronautics and Space Administration (NASA/JPL).*

Figure 35. *Hurricane Gordon in the Atlantic Ocean in 2006. If current trends continue, the frequency of severe storms, which gain strength in warm weather conditions, is predicted to increase. Photograph by National Aeronautics and Space Administration.*

7. CONSEQUENCES OF PRESENT & FUTURE CLIMATE CHANGE

Climate change is not inherently "bad." Indeed, we could all imagine how climate change could benefit this or that aspect of human life, and there is no way to define "the best of all possible worlds."[61] Although, for example, a warming climate could potentially be seen as positive for cooler regions – for instance, increasing the agricultural growing season, decreasing heating expenses, etc. – climate change is not merely a slight warming of the planet. As we have mentioned several times elsewhere in this book, because everything within the climate system is connected, changes in temperature beget changes in precipitation, humidity, ocean circulation, wind patterns, and many other aspects of the Earth's climate system. Therefore, assessing the consequences of present and future climate change is more complex than it might at first appear.[62]

7.1 Temperature

Based on the greenhouse gases that humans have already released into the atmosphere, and those predicted to be released based on current

Table 4. *Three emissions scenarios from the IPCC 2007 report. "Best estimate" refers to change in average global temperature of 2090-2099 relative to 1980-1999. "Likely range" refers to a range of values from a variety of models representing these scenarios (see Figure 29). Modified after IPCC, 2007 (see endnote 7).*

	approx. CO_2 concentration in 2100	**best estimate**	**likely range**
B1: likely best case scenario: world population peaks, energy-efficient technologies, and global cooperation	600 ppm	1.6°C	1.1-2.9°C
A1B: middle scenario: world population peaks, energy-efficient technologies, mix of fossil and non-fossil fuels	850 ppm	2.8°C	1.7-4.4°C
A1FI: business-as-usual scenario: continuous population growth, less global cooperation, mostly fossil fuels	1,550 ppm	4.0°C	2.4-6.4°C

trends, the most recent IPCC report predicts a 0.1° to 0.2°C warming each decade for the next two decades.[63] In 1990, IPCC scientists predicted a 0.15° to 0.3°C increase per decade for 1990 through 2005.[64] During that time, temperature actually increased by approximately 0.2°C per decade, which strengthens confidence in their projections. Thus, even without any increasing greenhouse gas emissions over that which occurred in year 2000, this warming of approximately 0.1°C will likely occur. It is also likely that continuing to emit greenhouse gases at or above 2000 rates will induce larger changes than those seen over the last century.[65]

Based on the relationship of different aspects of the climate system, such as ocean and wind patterns, the greatest warming is expected to be seen over land, and in high, northern latitudes. The least amount of warming is predicted to be seen in the Southern Ocean around Antarctica, and along parts of the northern Atlantic Ocean.

7.2 Storms and Other Severe Weather

Although it is impossible to link an individual storm event or season of storm events to climate change, it is not impossible to explain why they could be linked. Because climate is a complex system averaged over many years, the only way to unequivocally correlate severe weather events, such as hurricanes and cyclones, with climate change is to watch these events occur over at least several decades, average their intensities, and compare the climate and event records. A year or two of data, in other words, is not sufficient to causally link trends in weather events to climate change.

The science of severe weather events has progressed to the point that many basic aspects are well understood, and do in fact suggest a link to climate change. For example, conditions that create violent thunderstorms and tornadoes often occur on extremely hot days, and monsoons and hurricanes need very warm ocean waters to gather strength. A hypothesis that hurricanes and tornados will increase in frequency and/or intensity as the climate warms stems from the logic that an increase in global temperature would increase the number of very hot days, and would continue to warm ocean waters. There might also be mitigating factors, however, and a scientific consensus on the issue has not yet emerged using data from the 20th century; the hypothesis will continue to be tested over the coming decades.[66]

7.3 Ground Water

Humans have built their lives and civilizations around access to water. Water is used to fulfill basic human needs, including drinking, cooking, and bathing, and also transportation and agriculture. It is a resource that is taken for granted in places where water resources are abundant, and hotly contested in places where it is in short supply.

In general, climate models suggest that water will become harder to obtain; for example, the dryer latitudes – the belts at approximately 20° latitude north and south that contain many of the world's largest deserts – will likely expand. Even in the northeastern U.S. in coming years, residents will probably experience increased scarcity of freshwater (see Appendix 5). More hot days will increase the likelihood of more violent weather, such as thunderstorms, hail storms, and localized tornadoes, all of which will increase chances of supersaturating and flooding land, and not allow the water to slowly seep into groundwater aquifers. Decreased summer precipitation will increase droughts, damaging crops, wildlife, and recreational activities. Furthermore, hundreds of millions of people around the world (in California and Bangladesh, for example) depend upon glacial meltwater from nearby mountains, and these glaciers are shrinking. In summary, although total precipitation might not change much over many areas, the combined effects of episodic extreme precipitation and higher temperatures will likely decrease soil moisture and groundwater tables, and therefore, the supplies of water available to large numbers of people.

7.4 Sea Level

Counterintuitively, the melting of sea ice (which is different from the continental ice sheets that are not currently displacing any water) has no appreciable effect on sea level because it is already floating on the surface, displacing water; when that ice melts, the solid turns to liquid but the same amount of space is occupied. So, how can sea-level rise as a result of global warming (Figure 36)? First, water has the property of expanding when heated. Therefore, as the ocean water warms, it causes sea level to rise just by being warmer. A global temperature rise of 1°C corresponds to approximately 3.3 meters (10 feet) of sea-level rise; because the oceans take a long time to heat up, such a rise would occur over the course of several centuries.

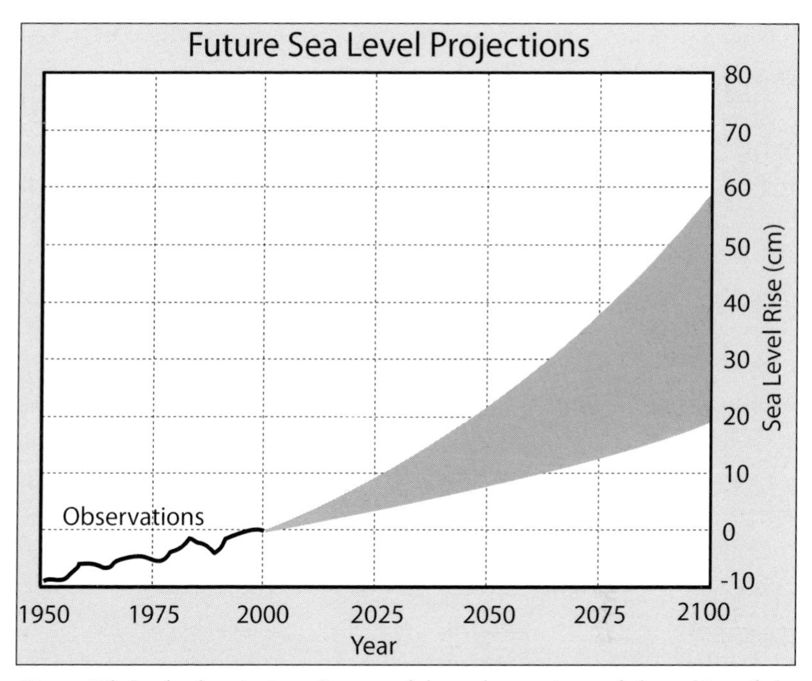

Figure 36. *Sea level projections. Because of thermal expansion and the melting of glaciers, sea level is predicted to rise as temperature rises. Based on various emission scenarios (see Chapters 6 and 7), a sea-level increase of 20-60 centimeters during the next 100 years is predicted. Modified after a graph produced by Robert A. Rohde for Global Warming Art.com.*

In contrast to sea ice, continental ice sheets sit atop land and do not already displace water in the ocean. Thus, when the continental ice sheets melt, water flows into the ocean and sea level rises. The Greenland and Antarctic ice sheets are the two largest ice sheets in the world, holding approximately 2.85 million cubic kilometers and 30 million cubic kilometers of ice, respectively. If the Greenland ice sheet were to melt entirely, sea level could rise by approximately 7.3 meters, or 24 feet. If the entire Antarctic ice sheet were to melt completely, sea level could rise by approximately 70 meters (230 feet). Although climate scientists are not predicting that either of these ice sheets will completely melt anytime soon, it points to the potential magnitude of future changes.

Like other aspects of climate change, sea-level change is not a new phenomenon. As recently as 20,000 to 10,000 years ago, the end of

Earth's last glacial maximum, sea level rose approximately 110-120 meters (350 feet). Similar changes have occurred throughout the geologic record. At various times of glacial development, sea level dropped because a tremendous amount of water was frozen in ice sheets. As the Earth began to warm, these glaciers melted and sea level began to rise. Sea-level change on even larger scales has occurred as ocean basins changed shape and size due to plate tectonics, the process that moves entire continents over hundreds of millions of years (see Section 2.4.3).

By approximately 10,000 years ago, much of the glacial melting was complete, and records show relatively stable sea levels during the development of major human civilization since then. This is important because many of the world's great cities are located along the coasts, mainly due to availability of resources and access to global trade. These cities could be devastated by a major future sea-level rise.

Current predictions of sea-level rise are intricately dependent upon predictions of and interactions between temperature change and the melting of polar ice caps. This means that specifying the expected levels of sea-level change over the next 100 years is very difficult. In 2001, the IPCC predicted an increase of 0.1-0.7 meters (0.5-2.5 feet) by the year 2100.[67] Maps reflecting such a rise show that a one-meter (three-foot) rise in sea level would swamp cities all along the eastern U. S. seaboard. In sumary, sea-level rise is expected to continue, but its total rise and the rate at which it is expected to rise are dependent upon other aspects of climate change and human CO_2 emission rates.

7.5 Agriculture

As already mentioned, climate change will likely cause precipitation patterns to change, and severe storms will become more frequent. This, in addition to a warming climate, will dramatically affect how farmers are able to grow crops. Both intense rainfall and drought are, in general, bad for agriculture, because heavy rains drown crops, and drought starves them of water. Both of these situations have been predicted to increase with climate change. Also, farmers in each region are accustomed to growing a range of select crops that thrive in their particular temperature range. As the Earth warms, these ranges will change, forcing many farmers to change their crops. Here are some examples of how agriculture in just one region – the northeastern U.S. – could be affected.

Particular crops: Climate change could mean that many crops that have been grown in a region for a long time will no longer be able to be grown there. Although they might be able to be grown elsewhere, many such crops have become tightly interwoven into the local economy as well as the culture and identity of the local population, so their disappearance would therefore have a spectrum of disruptive effects. For example, sugar maple trees, and the associated syrup industry, have been a characteristic part of rural life in New England for centuries. Earlier, warmer springs, however, could lessen the amount of maple sap extracted during the sugaring season. Some scientists have even suggested that within 100-200 years, sugar maples could completely disappear from the northeastern U.S.[68]

Apples are another example of a northeastern U.S. crop that could be impacted negatively by climate change. Although a longer growing season might at first seem positive, many apple trees need a certain number of days below freezing to set a large amount of fruit. With warmer winters, many traditional varieties of apples will no longer produce large amounts of big fruit, and with warmer and earlier springs, apples will bloom earlier. Spring temperatures are very changeable, and if a bloom occurs early, followed by frost, the flowers and fruit could be damaged. Increased climate variability is particularly problematic for perennial crops, such as fruit trees and vineyards, because adjustments cannot easily be made to long-term investments in the crops and specialized equipment.

Crop yields: Increased atmospheric CO_2, as well as a longer growing season, could boost yields of some crops. However, higher ozone concentrations associated with unusually hot days, particularly near urban centers, can damage soybeans and other crops, countering positive impacts of a warmer climate. In addition, severe storms and floods during the planting and harvest seasons could decrease crop productivity. Hotter and drier summers and potentially more droughts would hurt crops and could require irrigation of previously rain-fed crops, increasing the costs for farmers as well as the pressures on water resources.

Livestock: Higher temperatures suppress appetite and decrease weight gain in many livestock species, whereas warmer winters and less snow cover could reduce the quantity and quality of spring forage, decreasing, for example, milk quality. In addition, extreme weather events, such as heat waves, droughts, and blizzards, can have severe effects on livestock health, although intensively managed livestock operations are better able to buffer the effects of extreme events.

Pests: Warmer winters – with longer freeze-free periods, shifts in rainfall, and extended growing seasons – could create more favorable conditions for many agricultural pests. More southerly pests could expand northward; such shifts already appear to be happening in the northeastern U.S. with beanleaf beetles and corn earworms. Warming will increase the rate of insect larval development and the number of generations that can be completed each year, contributing to an increase of pest populations. Increased pests might drive farmers to use more pesticides, placing an additional burden on water quality and human health.

Soil erosion: Heavy rains and flooding could lead to an increase in farmers' costs to maintain soil fertility as well as contribute to off-site costs, including nutrient overloads and pollution to local waterways.

7.6 Human Health

Human populations on Earth have always been directly impacted by climate. Ancient Egyptian, Mayan, and European civilizations experienced growth or decline based in part on climate cycles. An outbreak of disease, for example, often occurred in response to climate changes that brought extremes in weather or storm activity. Therefore, as the Earth enters this new phase of climate change, scientists expect to see some dramatic challenges to human health.[69]

Heat: With current climate change, the first obvious threat to human health is the warming itself. In New York State, for example, the number of days each year over 32°C (90°F) is projected to increase to 40° or more by the end of the 21st century. Temperatures exceeding 36°C (97°F) are considered "extreme heat," which is associated with heat exhaustion, heat stroke, cramps, and fainting. If heat waves last longer and recur more frequently, people will be increasingly vulnerable to these health problems.

Air quality: When temperatures are high, air quality frequently declines. Hot days can exacerbate the production of ozone, which occurs when nitrogen oxides and volatile compounds from tailpipe exhaust, industrial emissions, gasoline vapors, and chemical solvents interact in the presence of sunlight. Ozone, the primary component of smog, can make breathing difficult, cause lung irritation, wheezing, and coughing during outdoor activities. It can aggravate asthma, increase incidence of pneumonia and could ultimately cause permanent lung damage with repeated exposure.[70]

Infectious diseases: An increase in global average temperature will likely be accompanied by an increase in the incidence and spread of infectious diseases. Humans have long known that there is frequently a correlation between warm temperatures and disease. Prior to the advent of modern medicine and hygiene, for example, wealthy families would try to avoid malaria by leaving low-lying cities in the heat of the summer for the cooler highlands. Many people living in lower latitudes, such as South Asians and Latin Americans, learned that spicy foods helped ward off many diseases (today we know this is because such spices kill many disease-causing microbes). It is thus perhaps not surprising that a warmer world will likely see an increase in the frequency and rate of spread of many human infectious diseases, including malaria, cholera, dengue fever, West Nile virus, and Lyme disease. This is for at least two reasons. First, the microbes (bacteria and viruses) that cause many such diseases thrive in warmer temperatures. Second, they are often spread by other organisms – such as rodents or insects – which do better in warmer temperatures. Waterborne diseases, like those that cause diarrhea, are also likely to increase. Altogether, such "warm-weather" diseases today kill nearly 60 million people each year. This number can be expected to increase significantly as average global temperatures rise.

7.7 Coral Reefs

As mentioned in Chapter 3, most **corals** are colonial organisms that live in the sea and make their skeletons from calcium carbonate ($CaCO_3$). Coral reefs, which are large structures on the sea floor built from the skeletons of many corals as well as other marine organisms, are extremely important elements of the marine environment. They are home to thousands of species of organisms, more than any other type of habitat in the oceans, and are also major components of larger coastal ecosystems, protecting shorelines from the effects of storms and providing habitat for fish that in turn provide food to many other animals, including humans.

Coral reefs around the world are currently in extremely poor shape and getting worse, and most scientists have concluded that this is at least partly to do with the effects of ongoing climate change. Most corals are extremely sensitive to temperature. An increase of just a few degrees can cause them serious problems. These problems are often evident in a phenomenon called **coral bleaching**, in which the corals become whitened and then usually die.[71] Bleaching is a complex phenomenon and probably has multiple causes, but warmer water and increased dissolved CO_2 in the water are almost certainly among them. Major bleaching events have already been associated with rises in ocean temperatures. A single warming event associated with El Nino in 1998, for example, resulted in the death of 16% of the world's reefs, and bleaching events on the world's largest reef, the Great Barrier Reef in Australia, have affected approximately 50% of the coral, killing off 2-5% of the coral per event. Numerous recent studies have concluded that predicted increases in atmospheric CO_2 (and associated environmental changes, including warming and ocean acidification; see below) would be catastrophic for most coral reefs. For example, if current trends continue, mass coral bleaching might happen every two years in the eastern Caribbean within 20-30 years. Even if corals are able to adapt to slightly warmer waters, such events are almost certain to become frequent by the latter half of the 21st century. Some marine biologists have even suggested that, if current trends continue, coral reefs could become virtually extinct worldwide by early in the next century.[72]

7.8 Ocean Acidification

A large proportion of recent increased CO_2 emissions have been absorbed by the oceans, which act as a natural buffer, or sink, for CO_2. As a result of this absorption, however, oceans have become more acidic (just as water becomes acidic when we "carbonate" it for soft drinks by bubbling CO_2 through it). It is estimated that the average oceanic pH has dropped by 0.1 unit since pre-industrial times.[73] Although it is known that local oceanic pH can vary dramatically throughout a day (up to 0.24 units) and throughout the year (up to 1.5 units), over the last several years, global oceanic pH has displayed a consistent downward trend toward more acidic ocean waters.[74] The oceans will continue to become more acidic as more CO_2 is absorbed. If current trends continue, it is predicted that the pH of the oceans will drop between 0.14 and 0.35 units.[75]

Normally the warming associated with additional CO_2 emissions is coupled with increased erosion and weathering, and the sediments created by that weathering find their way into the ocean and raise the pH, countering the acidification. Unfortunately, the creation of dams worldwide now captures approximately 20% of the sediment carried by rivers, and most of the sediment that reaches the oceans is from lowland farms, not high-elevation rocks that would produce sediments that could raise the pH. As a result, restoring the oceans' pH to their pre-industrial levels could take thousands to hundreds of thousands of years.[76]

Acid dissolves calcium carbonate ($CaCO_3$), the material that many marine organisms – including snails, clams, corals, sea urchins, crabs, lobsters, and some algae – use to build their skeletons. As a result, the shells of these organisms have become weaker, threatening their survival. Not all marine organisms, however, will respond the same way to these changes in ocean pH.[77] Research into this question is in its very early stages, but some studies suggest that shallow-water organisms, such as many kinds of mollusks, echinoderms, and algae, could be less sensitive than their deeper-water relatives. Other studies show much more complex results. One recent study, for example, found that conch shells (large, sturdy marine snails) deteriorate very rapidly in a more acidic environment, whereas some crustaceans, such as lobsters and blue crabs, grow heavier shells under similar conditions.[78] Ominously, some recent studies of the fossil record suggest that some episodes of mass extinction in the oceans, might have been triggered by increased ocean acidification.[79]

7.9 Biodiversity

Biodiversity is the variety of life forms found in a specific environment, be it a garden, a rainforest, or the entire planet. Biodiversity is most simply measured as the number of species, and declines in biodiversity usually reflect the declining health of an environment or biological system. Just like climate, the Earth's diversity has changed dramatically over time. For example, more than 99% of all the species that have ever existed on Earth are now extinct. Sometimes dramatic numbers of extinctions have occurred during a relatively short interval of geologic time. Such episodes are called **mass extinctions**. For example, dinosaurs dominated the world for over 100 million years, but all (except the birds) went extinct approximately 65 million years ago. Mass extinctions can happen for many reasons, including climate change, continental position, sea-level change, extraterrestrial impact, or a combination of these.[80]

Biodiversity is important for human welfare for many reasons.[81] First, the health of an ecosystem can often be measured by a change in the number of different species found in it; a decline in diversity might mean that the entire ecosystem is declining. Second, wild species provide us with a huge array of products, foods, and chemicals that are important for our health, economic prosperity, and comfort. Most modern prescription drugs, for example, are based on chemical compounds originally found in tropical rainforest plants. Many domesticated crops depend on periodic infusions of genetic diversity from their wild relatives. Healthy natural ecosystems have enormous effects on weather, soil erosion, water supplies, and recreational resources.

Human impact on biodiversity is not limited to climate change. Overexploitation, pollution, invasive species, and habitat alteration (deforestation, urban sprawl, agriculture) have already caused the ex-

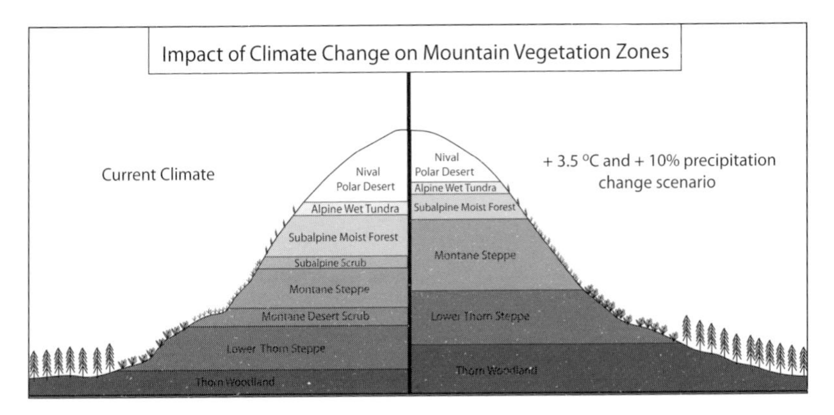

Figure 34. *Mountain ecosystems are a good example of how habitat loss can occur due to climate change. As global temperature increases, mountain slopes with unique, temperature-sensitive ecosystems are forced upward toward the top of the mountain. Although some habitat expands at the base of the mountain, other habitats shrink, or even disappear as a result of a warming climate. Modified after a graph from the UNEP/GRID Arendal Maps and Graphics Library.*[82]

tinction of dozens, perhaps hundreds, of species, especially in tropical rainforests and coral reefs. Many biologists argue that we are in the midst of a major mass extinction of the magnitude of the one that saw the demise of the dinosaurs.[83]

Global climate change is just one more pressure being put on Earth's biodiversity by humans, and for many species it might be the straw that breaks the camel's back.[84] Rising global temperatures will have many effects on species. At particular risk, of course, are the species that do not thrive at temperatures above those that they inhabit today. If they cannot move to cooler areas, they will disappear. This appears to be happening with some species of small mammals and amphibians living on isolated high mountain peaks (Figure 37). Even if a species can move, however, such migration could itself cause problems for other species. For example, many species of birds and insects have already been observed to be shifting their geographic ranges toward higher latitudes. In doing so, they can compete with other species already living there, causing one or both to decline in abundance. These climate-caused range shifts, furthermore, are taking place on landscapes that are already fragmented by other human activities, such as agriculture, deforestation, roads and other development, and so species might not be able to move as freely into new habitats as they might otherwise. Just as for humans, the incidence and spread of many diseases affecting wild plants and animals could also increase with warming.[85] In the oceans, rising water temperatures and acidification are already having marked effects on the health of reefs and the thousands of species that they contain. Changing climates are also making it easier for many invasive species – from kudzu to zebra mussels – to enter new habitats, displacing native species.

Summary: Consequences of Present & Future Climate Change

Anthropogenic climate change will have numerous, major negative consequences that will significantly impact humanity across the globe. The most obvious are the negative consequences associated with globally higher temperatures, the necessity of changing agricultural practices, and increased storm frequency and severity. More subtle, but perhaps of even greater consequence, are the negative impacts on oceans, with rising acidity levels and rising waters, weakening and die-off of coral reefs and associated ecosystems, and overall biodiversity loss. Climate change will also have negative impacts on human health and current standards of living as freshwater resources become more scarce. Some of these negative consequences are already apparent, and it is very likely that they will increase as greenhouse gas emissions rise.

Figure 38. *The Golden Toad* (Bufo periglenes) *was last seen in Costa Rica's Monteverde Cloud Forest Preserve in 1989. Its decline and ultimate disappearance was largely blamed on exceptionally dry, El Niño-driven conditions that eliminated the forest pools necessary for survival of the toad's tadpoles. Following extensive searches, it was declared extinct in 1999, marking the first documented species extinction driven by climate change. Photograph by U.S. Fish & Wildlife Service.*

Figure 39. *Scientists prepare to drill a sediment core from a raft on a lake in eastern Greenland to recover sediments for the reconstruction of paleoclimate and the glaciation history of Greenland. Photograph by Hannes Grobe, Alfred Wegener Institute for Polar and Marine Research, via Wikimedia Commons.*

8. Science & Certainty

8.1 How Does Science Work?

The idea that the Earth's climate is changing as a result of human activity is not just an opinion, or a guess; it is a scientific conclusion. **Science** is an approach to explaining and understanding the natural ("material" or "physical") world. Its philosophical approach is frequently called materialism or naturalism. It uses observations about the world and the rules of logic to test hypotheses that explain natural phenomena. **Hypotheses** are ideas about natural phenomena; they might or might not be true. Hypotheses can come from anywhere. What makes a hypothesis scientific is that it is testable. Testing hypotheses means making predictions from them, and then comparing these predictions to observations from the physical world. Hypotheses that pass such tests are accepted, but such acceptance is always provisional, that is, any accepted hypothesis can be overturned by sufficient, credible contrary evidence. A **theory** in science is an idea or set of ideas and hypotheses that connects, explains, and is supported by a large number of observations. Although in common English, a "theory" can mean a mere guess or supposition, in science, it is the basic unit of our understanding of reality; a theory in science is about as good as it gets.

The main technique that scientists use to put forward and test hypotheses is **peer-reviewed literature**. Here is how it works: When a scientist has an idea, and at least some evidence that appears to support it, he or she writes a manuscript explaining the idea and the evidence, and where the evidence came from. The manuscript also includes references to previous scientists' work to provide context and to summarize the evidence and arguments that already exist. Sometimes two or more scientists will collaborate to write a manuscript together. Once a draft of the manuscript is complete, the author(s) might present the paper at a conference, and/or might ask one or more colleagues to read and comment on it informally, and either or both of these could result in changes. Once a "final" draft is ready, it is submitted to a scientific journal. The editor of the journal (usually an expert in the general field covered by the journal) reads the submitted manuscript, and then sends it to one or more other experts who he or she thinks have expertise in the subject. These readers are called peer reviewers because they are professional "peers" of the author(s). Reviewers are asked to look particularly at whether the authors' data appear to be valid and trustworthy, whether the data support the conclusions, whether alternative conclusions are equally likely, and whether the logic, writing, and presentation are clear and convincing. After reading the manuscript, each reviewer writes formal (sometimes lengthy) comments back to the editor. Depending on the policies of the particular journal, the editor might keep the identity of the reviewer(s) anonymous to the author, or might reveal their names. Anonymous reviews are generally seen as likely to be more honest, and are used by the majority of scientific journals.

Based on the reviews and his or her own reading of the manuscript, the editor writes to the author, usually with one of four responses: (1) the manuscript can be accepted for publication in the journal without any revision (this happens very rarely); (2) it can be accepted only if relatively minor revisions are made by the author(s); (3) it can be accepted only if major revisions are made; or (4) it is rejected and will not be published by the journal. The editor sends copies of the reviewers' comments – both to support the response about acceptance and to assist the author in revisions. Depending on this response from the editor, the author then revises and resubmits the manuscript to the same journal, or starts over with another journal. Some manuscripts repeat this process many times before they are finally accepted for publication. Some manuscripts are never published. Not all journals are equal in reputation or prestige. Some – such as *Science* and *Proceedings of the National Academy of Sciences* in the U.S., or *Nature* and *Proceedings of the Royal Society* in the U.K. – are extremely selective in what they accept for publication. These journals are therefore viewed by authors as the most elite and sought-after places to publish, and by readers as among the most reliable sources of accepted conclusions. Scientists are not paid by the journal in which their paper appears. In fact, many journals charge authors to publish in them.

The peer-review literature system is critical to the professional life of every scientist. Getting manuscripts published as scientific papers in peer-reviewed journals is the basis for getting and keeping jobs and research grants. (Publishing in non-peer-reviewed outlets – such as popular magazines, books, or (increasingly) blogs or websites – is not an acceptable substitute in most circumstances.) Authors, reviewers, and editors therefore take this process extremely seriously. Reading critical reviews of a submitted paper is something that every scientist has experienced, and it can be one of the most humbling (and sometimes infuriating) experiences in a scientist's career.

Like all human activities, peer-review is not perfect. Some reviewers base their comments not just on the content of the manuscript but also on the identity or reputation of the author, or whether the paper supports their own views. Anonymity protects honesty but can also protect personal biases and the exclusion of unpopular or controversial conclusions. Editors are supposed to sort this out, but sometimes they do not. Established, senior scientists at well-known institutions are often judged by different (sometimes more relaxed) standards than younger researchers, or those from lesser-known institutions. Fads or fashions can sweep through fields, temporarily lowering standards, and every journal at least occasionally publishes a paper that is later seen as not measuring up to the journal's usual standards. All scientists want to publish in the most prestigious journals, but there are in fact so many scientific journals that almost any paper can eventually be published somewhere. Although it should be possible in principle to reproduce or at least verify all of the data reported in any paper, in practice there are few attempts to intentionally replicate most experiments or observations unless there is a particular reason to do so (such as anomalies and exceptions that arise as side-consequences of other work). Presumptions of trust and honesty therefore play central roles in the system. Very occasionally, there is outright fraud, such as the publication of fabricated or manipulated data. Yet, despite its flaws, the peer-review system by and large works, meaning that it produces results that most knowledgeable scientists in the end view as the best so far available.

What scientists provisionally accept as "true" is largely a result of this peer-review process. A scientific **consensus** reflects the general understanding of an issue held by the great majority of specialists who work on that issue, as expressed in the peer-reviewed literature. The degree of consensus among specialists who have studied a subject for a long time is some measure of the confidence with which scientists in other specialties, as well as nonscientists, can accept a scientific conclusion. For example, if more than 90% of papers on a particular topic published in the peer-reviewed literature during the past two years are in agreement on a particular conclusion, or at least aspects of it, it is generally accepted with a high degree of confidence. If, on the other hand, there are roughly equal proportions of papers on either side, or multiple competing hypotheses in the recent literature, no particular conclusion could be said to be accepted with high confidence.

For the purposes of this book, two major points about this definition of scientific consensus are important to note:

- **Consensus among specialists:** Whether a conclusion is broadly accepted by scientists is not based upon a simple vote of opinions; it is supposed to be decided by agreement of hypotheses and theories with observations about nature. In practice, however, this means that what is accepted as true in a particular area at a particular time by the overall scientific community is usually the majority view among those scientists who are specialists in that area. If most specialists who have devoted years to researching a topic accept (provisionally) a particular theory (*i.e.,* if there is a consensus), it is usually treated as "true" by other scientists who have not studied it in great detail. This means that what is accepted is not necessarily what *all* specialists think, nor is it what the majority of *nonspecialists* think. Not all cancer specialists, for example, might agree on the cause of a particular form of the disease, but if a clear majority do, then this is taken as at least temporarily indicative of the actual cause. In such a case, however, it is probably irrelevant what the majority of doctors in other, seemingly unrelated medical specialties (such as orthopedics or podiatry) think is the cause. (It is certainly not going to matter what a majority of paleontologists think!)

- **Certainty, and the role of uncertainty:** Science is not about absolute certainty – proving or knowing that something is true without any possibility of doubt or error. Because all scientific conclusions – even very well-supported ones accepted by most or all specialists – are provisional, they could be incorrect, and this could (and presumably would) eventually be demonstrated by discovery of sufficient, valid contrary data. Thus, although many scientific conclusions are so well-supported by so much evidence that they are treated as effectively certain, *no scientific conclusion is absolutely certain*. Rather, scientists accept them with various degrees or certainty or confidence, based on the data and arguments available, as presented in the peer-reviewed literature. This means that the existence of some amount of uncertainty about a scientific conclusion is not necessarily an indication that that conclusion is incorrect. Furthermore, even considerable uncertainty (*i.e.*, lack of consensus) about one aspect of a conclusion might have no bearing on the confidence with which other aspects are accepted.

8.2 The Current Consensus on Climate Change

The IPCC reports represent a scientific consensus view of climate change. The IPCC itself includes more than 4,000 thousand climate scientists from nations around the world. In addition, dozens of other statements have been put forward by scientific organizations around the world, representing tens of thousands more scientists, supporting the consensus position represented by the IPCC reports.[86] As described above, this does not mean that the conclusions expressed by the IPCC are all correct. It means only that the great majority of climate scientists accept most of the IPCC conclusions. There is no comprehensive tally, but a recent poll found that more than 97% of active **climatologists** – scientists who specifically work on climate research – accept the conclusions that global temperatures are rising and that most or all of this increase is due to human activity.[87] "Consensus as strong as the one that has developed around this topic is rare in science" wrote *Science* magazine's executive editor Donald Kennedy in a 2001 editorial.[88]

This does not mean that all climate scientists accept all of the IPCC conclusions, or have the same degree of confidence in each of them. Although there is extremely widespread agreement on a few central conclusions – global temperatures are rising, that most or all of this increase is due to human CO_2 emissions, and that the consequences could be quite dangerous for humans – there remains considerable uncertainty about many other peripheral conclusions, such as whether hurricane frequency and intensity will increase with rising temperatures, what the role of clouds and other particulates in the atmosphere is, how much CO_2 will be absorbed by the oceans, and where "tipping points" – beyond which changes cannot be reversed – might be. Disagreement about these topics, however, has essentially no bearing on the widespread agreement on the larger conclusions. Authors of the

Box 11: Certainty and Confidence in the IPCC Report

?

The 2007 IPCC report[89] classified scientific uncertainties into two categories – "value uncertainties" and "structural uncertainties." Value uncertainties were defined as those that come from the incomplete determination of particular values or results, for example, when data are inaccurate or not fully representative of the phenomenon of interest. Structural uncertainties come from an incomplete understanding of the processes that control particular values or results, for example, when a hypothesis or model used for analysis does not include all the relevant processes or relationships. The report noted that value uncertainties are generally estimated using statistical techniques and expressed in terms of mathematical probability, whereas structural uncertainties are generally described by summarizing the relevant specialists' "collective judgment of their confidence in the correctness of a result." In both cases, the report stated that "estimating uncertainties is intrinsically about describing the limits to knowledge and for this reason involves expert [*i.e.,* informed but subjective] judgment about the state of that knowledge."

The IPCC report also for the first time made a careful distinction between levels of confidence in scientific understanding and the likelihoods of specific results. This allows authors to express high confidence that an event is extremely unlikely (*e.g.,* rolling dice twice and getting a six both times), as well as high confidence that an event is about as likely as not (*e.g.,* a tossed coin coming up heads). The terms used in the report for these certainties (as a percentage of probability of occurrence) were:

Virtually certain	> 99%
Extremely likely	> 95%
Very likely	> 90%
Likely	> 66%
More likely than not	> 50%
Unlikely	< 33%
Very unlikely	< 10%

The IPCC also used the following levels of confidence to express expert judgments on the correctness of the underlying science:

Very high confidence: having a 9 in 10 chance (or better) of being correct
High confidence: an 8 in 10 chance
Medium confidence: a 5 in 10 chance
Low confidence: a 2 in 10 chance
Very low confidence: having a 1 in 10 chance (or less) of being correct

2007 IPCC report tried to deal with these issues by explicitly defining levels or ranges of uncertainty, confidence, and likelihood for its conclusions (Box 11).

Despite the careful qualifiers of confidence and uncertainty summarized in Box 11, and the wide and growing breadth of the scientific consensus, the public – and even many scientists outside of climatology – remains confused about the scientific basis for various statements about climate change. This confusion has had major political, economic, and social implications. Yet the sources of the confusion are themselves confusing and complicated. If the majority of scientific specialists are so convinced that climate change is occurring and humans are largely responsible, why does it remain so controversial?

This is a large and complex question.[90] We will only highlight a few of the most important and obvious answers.

- **"Is" versus "ought."** The conclusions that the climate is changing, that humans are responsible, and that these changes will be harmful to a large proportion of the population, are not the same thing as deciding what (if anything) to do about it. Although science is not isolated from the larger society and it cannot conscientiously announce its findings with no sense of their political, social, or economic implications, the primary responsibility of scientists is to discover how the world is, not what human beings, individually or as a society, should do as a result of the world being that way. Making these decisions is ultimately the responsibility of politicians and policy makers.

- **The pace and nature of science versus policy-making.** Science usually works much more slowly than what is expected by the public and policy makers when they want information or decisions. Depending as it does upon the results of observations and experiments, which might be ambiguous or difficult to execute, science cannot usually be rushed to meet deadlines. Government and the general public usually want "answers" when they want them, not when scientists finally feel confident enough to state their conclusions. The resistance of scientists to rush to conclusions prematurely can be misinterpreted by nonscientists as doubletalk or lack of the ability to reach any conclusion at all. For their part, scientists frequently worry too much about their own confidence in the details, without realizing that more general conclusions (about which they have much greater confidence) might be more important.

- **Use and abuse of uncertainty.** If you think that science is about absolute certainty, and then hear that scientists are uncertain about a particular idea, then you might think that the idea itself is not correct. Given that a large proportion of the general public (at least in the U.S.) mistakenly thinks that this is what science is about,[91] opponents of a particular scientific conclusion can cause considerable confusion simply by emphasizing that some uncertainty exists about one or more aspects of that conclusion – even if that uncertainty is small, held by a very small number of competent researchers, or is limited to only peripheral issues. Some critics of the consensus view on climate change have done exactly this.[92]

- **Role of the media.** The media plays an important role in forming public understanding of climate change.[93] One of the central aspects of this role is the issue of "balance." The usual hope of a journalist is to remain unbiased, and therefore be able to present, with equal credibility, both sides of an issue. On climate change, however, the media frequently tries to achieve this by "balancing" the views of thousands of credible researchers, like IPCC scientists, with arguments from lobbyists or politicians, or with a single researcher's unverified claims. This is not balance; it is equating personal opinion with scientific conclusions. At some point, the preponderance of scientific evidence has accumulated to a point where responsible reporters should give the scientific consensus much greater weight than dissenting claims of a few challenging the mainstream scientific conclusions.[94]

  Journalists are also driven to find a story that engages their audience, and stories highlighting the status quo are usually of much less interest than are controversies and paradigm shifts. So the media frequently reports a "scientific controversy" when one does not really exist. In this context, it is noteworthy that some climate change skeptics (see below) claim that many journalists covering science have an interest in promoting acceptance of global warming, regardless of its scientific basis. It is true that journalists do have an interest in promoting themselves and their writing, and that their employers want to boost their audience and sell advertising. But challenging the mainstream view, as climate skeptics do, actually *gains* journalists and their publishers far more publicity than promoting the established, consensus view. This might explain why so many sections of the media continue to publish or broadcast the claims of climate change skeptics, regardless of their merit.

- **Climate change skeptics.** Although the overwhelming majority of climate scientists accept that global climate is changing and that humans are responsible, the general public continues to perceive that the scientific community is still actively arguing these questions (see Section 8.3). This confusion is, in part, the result of the activity of a diverse group that are often collectively referred to as "climate change skeptics." They include some climate and Earth scientists, but also a larger group of scientists whose specialties are well beyond these fields, as well as various nonscientists such as former astronauts, lobbyists, and political operatives.

  There *are* climate and Earth scientists who genuinely doubt either that the climate is changing in a long-term way, or that humans are responsible. Most of the "climate change skeptic" community, however, is composed of nonscientists who have little first-hand experience with climate science, but have a variety of (mainly nonscientific) objections to the implications of the consensus that humans are causing climate change. These skeptics put forth a variety of arguments, such as the following:

  - ***Problems with data.*** These range from objections about the location of temperature recording sites (skeptics claim that they are too close to urban areas and their associated "heat islands"), to claims that average global temperatures have actually been falling (based on only the last few years, disregarding the preceding century or more of warming). There is some truth to the difficulty of turning hundreds of thousands of measurements, taken in many different ways and over a span of more than a century, into a single globally averaged dataset, and it is also true that differing weather station locations mean that it is probably impossible to arrive at a meaningful figure for actual global average surface temperature. However, neither of these points alter the conclusion that the overall trend of temperature measurements is toward warming. Furthermore, independent data, such as satellite measurements of temperature of the atmosphere, all agree that the Earth is warming.[95]

- ***Problems with interpretation.*** Skeptics argue that, even if the data are valid, science has simply not yet demonstrated a long enough trend in global temperature increase to conclude with high confidence that the warming trend is unusual. Even if there is warming, some argue that there is insufficient evidence supporting anthropogenic CO_2 as its primary driver (*e.g.,* the Sun might be the major cause). Finally, some argue that even if humans are responsible, there is insufficient evidence that the consequences of warming will be as dire as many climate scientists predict. These objections have been addressed throughout this book.[96]

- ***Climate scientists foster alarmism about climate change to boost their funding.*** This is not only demonstrably untrue, but also implausible. Although it is true that the U.S. government spends millions of dollars on climate research each year, and that this increased by 55% from 1994 to 2004, it is important to note that an increasing portion of this is spent on mitigation technology (ways of reducing emissions or the results of emissions) rather than pure research. Climate scientists point out that if they were after a bigger chunk of that money, their best bet would be to stress the uncertainties of climate change and call for more research, rather than call for action.

- ***Climate scientists' dependence on government funding ensures that they toe the official line of the consensus, thereby inflating it.*** Scientific research does indeed sometimes move in directions motivated by government, mostly by the availability of funding. But overall, science follows its own lead. Under President George W. Bush's administration, for example, many scientists claimed that they had been pressured to tone down findings relating to climate change. In fact, those campaigning for action to prevent further warming have had to battle against large vested interests, including the fossil-fuel industry and its political allies. Many of the individuals and organizations challenging the idea of global warming have received funding from such sources. This in itself does not necessarily mean that the skeptics are wrong, of course, but it does raise at least the possibility that these individuals and organizations are not acting completely independently of their interests. For further discussion of corporate funding of climate skeptics, see Mooney (2005; see endnote 92) and (2007; see endnote 66).

- ***Climate-change skeptics cannot get their findings published in the scientific literature.*** It has taken more than a century to reach the current scientific consensus on climate change. It has come about through a steadily growing body of evidence from many different sources, and the process has hardly been secret. Now that there is a consensus, those whose findings challenge the orthodoxy are always going have a tougher time convincing their peers, as in any field of science. But findings or ideas that clash with the idea of human-induced global warming have not been suppressed or ignored – far from it. In fact, many of the better arguments seized upon by skeptics have been based on contradictory findings published in prominent journals.[97] Many such objections to the consensus view of climate change have been discussed in previous chapters (see, *e.g.*, Chapter 4).

8.3 What Does the American Public Know About Climate Change?

According to recent polls, knowledge about climate change among Americans is decidedly mixed (Table 5). On the one hand, most respondents say that they do believe that global temperatures have risen

Table 5. *Summary of recent U.S. national opinion polls regarding aspects of climate change (n/a = not available). See text for data sources.*

Statement/Opinion	% in agreement 2006	% in agreement 2007	% in agreement 2009
Earth's temperatures have been rising for the last 100 years	n/a	84%	80%
Climate change is the biggest environmental problem of the world	n/a	33%	25%
Global warming dangers are generally exaggerated	30%	n/a	41%
There is solid evidence that the Earth is getting warmer	77%	n/a	57%
Global warming will pose a serious threat to my way of life in my lifetime	35%	n/a	40%

in the past 100 years. On the other, significant proportions of Americans believe that there is considerable scientific uncertainty about global warming, especially its causes, and that even if it occurs it will probably not be a significant problem that will affect them personally.

In the 2008 polls, 25% of respondents thought that global warming/climate change is the biggest environmental problem of the world, down from 33% in 2007. Eighty percent thought that the world's temperature has slowly been rising, down from 84% in 2007, whereas 18% thought that it has not been happening, up from 13% in 2007.

The personal importance of global warming to people changed little between 2006 and 2008, with 16-18% feeling that it is extremely important, 30-34% feeling that it is very important, and 30-32% feeling that it is somewhat important.[98]

Polls have also asked how the U.S. should take action on global warming. In 2008, 68% wished the U.S. to take action even if other countries do less, 18% favored taking action only if others do, and 13% favored no action. The populus was split on whether taking action before other countries would help or hurt the economy (33% help, 32% hurt), and also split on whether government (regulated)- or business (competitive)-promoted systems would do a better job of reducing greenhouse gas emissions (43% government, 45% business).[99] Polling by CNN and Gallup in 2008 showed that 49% wished for the environment to be given priority over economic growth, and 42-44% favored the reverse opinion.[100] Fifty-nine percent supported a cap-and-trade system, and 34% opposed it.[101] Despite mixed attitudes toward climate change, 71% of people indicated that they are personally doing things to reduce their carbon footprint, and 67% of people said that they were trying to buy products that are environmentally friendly, even though only 41% considered themselves environmen-

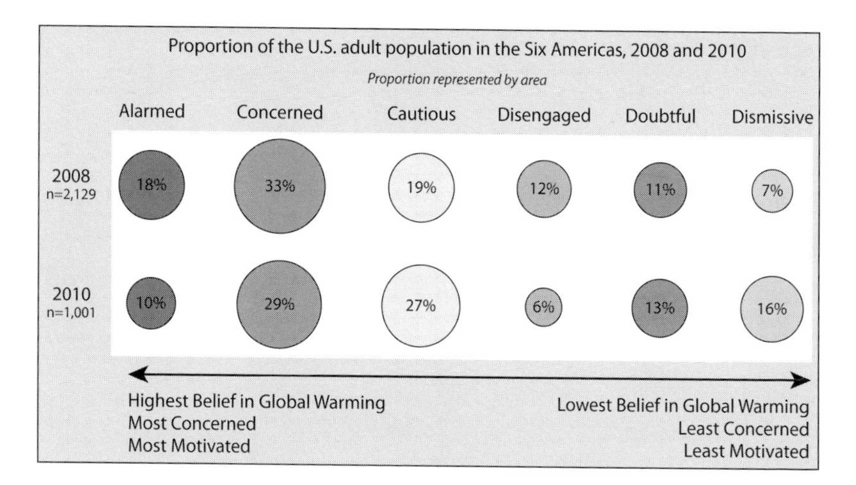

Figure 40. *Proportions of the U.S. adult population (2008 and 2010) in six categories of belief in global climate change. Modified after Leiserowitz,* et al., *2010: fig. 1 (see endnote 104); used with permission of the authors.*

talists. In 2008, only 40% (up from 35% in 2006 Gallup polling results) thought that global warming will pose a serious threat to their way of life during their lifetime.[102]

Since 2008, however, public opinion about climate change in the U.S. has changed significantly. In 2006, for example, a Gallup poll noted that more Americans thought that the dangers of global warming were underestimated than overestimated (38% versus 30%). But in a March 2009 poll, 41% of respondents said that the problem was exaggerated compared to 29% who said it was underestimated. In the same 2009 poll, 16% of respondents – the highest since the question was first asked in 1997 – said that they thought that the predicted effects of global warming would never happen.[103] An October 2009 Pew poll and December 2009-January 2010 poll by Yale University showed a similar attitude shift toward climate change.[104] In the Pew poll, only 57% of respondents said that they believed there was solid evidence that the Earth is getting warmer, down from 77% in 2006. In the Yale poll, 47% agreed that, assuming global warming is happening, it is caused mostly by humans, down from 57% in 2008. And only 34% agreed that "most scientists think global warming is happening," down from 47% in 2008.

Thus most Americans seem to accept that the world's temperature has been going up slowly over the past 100 years, but, despite the overwhelming scientific consensus, they are far from certain that humans are the major cause, or that it will end up mattering very much to them (Table 5). This pattern of public opinion has major implications for policy makers. If half of the public does not think that humans are the primary reason behind warming, then they will likely not support aggressive steps to do anything about it.

What is the explanation of these views? The recent revelations of emails between climate scientists that seemed to suggest suppression of contrary results,[105] and of errors in the 2007 IPCC report,[106] have encouraged critics of the scientific consensus. Yet this of course is not a general explanation for the remaining stubborn skepticism and doubt in the face of such overwhelming scientific agreement.

We are not sure that we understand why such resistance persists, but we have a few ideas.[107] First, people will frequently do everything that they can to avoid accepting unpleasant news.[108] Second, the lack of understanding of climate change mirrors a general lack of understanding science. This is not just a problem of lack of knowledge of "facts," such as that the Earth revolves around the Sun and is billions of years old, why hot air rises, that life evolves, why water dissolves solids, what the speed of light is, how many elements there are, that dinosaurs and humans did not live together, or how DNA works.[109] It is also the problem that many Americans do not seem to understand the nature of science, how science works, or why scientists know what they say they know. We have personally frequently been struck in particular by the widespread belief that science must be essentially unanimous in a conclusion – in a word, "certain" – before nonscientists should take it seriously.[110] This could not be farther from the reality of how science works, and it is a dangerous phenomenon when we are talking about issues of such potentially major magnitude as climate change.

Summary: Science & Certainty

To understand how climate scientists have come to the conclusion that anthropogenic climate change is occurring, one must also understand the nature of how scientific study is conducted, and that there is no "proof" in science. In fact, science is always trying to falsify itself to test the soundness of its hypotheses. The IPCC has instituted a scale of certainty to help the public understand exactly how well different facets of climate change are understood. Despite this, most people receive most of their climate change information from the media, which reports the science with varying degrees of accuracy. As a result, society has formed various opinions about what anthropogenic climate change is, how well scientists understand it, and what should be done personally and politically.

Figure 41. *President Barack Obama confers with European leaders at the United Nations Climate Change Conference in Copenhagen in 2009. Public opinion of climate change is largely shaped by media coverage of such events. Photograph by U. S. Federal Government.*

Figure 42. *TAKE THE BUS! The average American automobile gets only 12 miles per gallon in city driving, which translates into 24,500 pounds of carbon emitted per automobile every year. Photograph by The Port Authority of New York & New Jersey.*

9. WHAT CAN WE DO?

For the last eight chapters, we have explained the science behind the current climate change problem. That science tells us – with high and increasing levels of confidence – that current climate change is a serious threat to human welfare and that its primary cause is one of the central bases for that welfare: the burning of fossil fuels for energy. Thus, the best science that we have available tells us that modern society is built on some very unsustainable practices, and that, if current trends continue, those practices will put us all in a very difficult circumstance.

Science cannot tell us what we *should* do; it can only tell us what we *can* do, and how much we have to do, to achieve a desired result. In the case of climate change, science is telling us that we will very likely have to take some very large and very rapid steps very soon, and that the available options for such steps will all require a combination of individual and collective action. We must, in other words, change both our own behavior and that of our entire society.

The issue of "what to do about climate change" is thus one of the largest and most complex issues humanity has ever had to deal with. It involves how we produce and consume energy, and also how we might mitigate the effects of greenhouse gas emissions and adjust to the environment changes that are already unavoidable. We obviously cannot address all of this in detail here. What follows, therefore, is an overview of how each person can think about both his or her own personal actions and the actions of our larger society in dealing with this issue.

9.1 Your Carbon Footprint

A **carbon footprint** is the amount of CO_2 that an individual, household, company, or geopolitical area (such as a county, state, or nation) releases into the environment over a certain period of time. It is thus a simple measure of contribution to the primary cause of current climate change. Your personal carbon footprint is not just based on what kind of car or how much you drive, or on the energy efficiency of your home. It is also "hidden" in the goods you buy, how much you buy, and how you dispose of waste.

Many different carbon footprint calculators are available on the internet and in various books and magazines.[111] In most of these, the amount of CO_2 released is measured in tons, as we discussed in Chapter 2. Carbon footprint calculators generally measure an individual's carbon footprint in two ways: by measuring how much CO_2 an individual releases as a direct result of their actions, or by factoring in the CO_2-emitting infrastructural benefits that they receive by living in a certain region. Chapter 4, for example, discussed some contributing factors to anthropogenic greenhouse gas emissions other than fossil-fuel burning, including cement making and changing land usage. Because the land surface of industrialized nations such as the United States has been altered due to the increase in agricultural practices, and the construction of highways, skyscrapers, and cityscapes, and because these alterations on the land have contributed significant amounts to anthropogenic climate change, some carbon footprint calculators incorporate an individual's country into their calculations.

9.2 Energy Consumption

Humans – especially those in industrialized nations – use energy almost every minute of every day (Figure 42). We heat and cool and light our homes and workplaces. We travel by land, sea, or air. We build things. We make and consume food and other products that require energy for their production or manufacture. Most of the energy that we use to do all these things is produced by burning fossil fuels, and therefore adds CO_2 to the atmosphere. Thus, if we could reduce our consumption of energy, we might reduce greenhouse gas

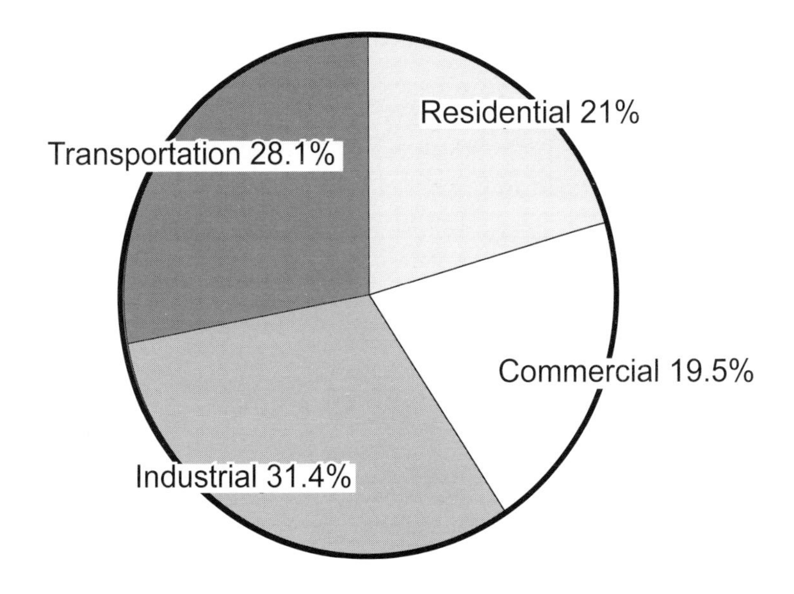

Figure 43. *Energy consumption in the United States.*[112]

emissions. For such reductions to have a significant effect, however, they must be implemented *both* by individuals and larger units of society. There will be no net long-term reduction in greenhouse gases, for example, if you walk to work instead of driving if the average fuel efficiency of cars sold in the U.S. goes down. Individual behavior matters, but so does the collective behavior of society.

Home: Nearly 21% of all energy used in the United States is used at home,[113] and there are many ways that individuals can increase home-energy efficiency. The first is to conduct an energy audit. The audit is intended to show you where you consume energy most and where energy losses exist. The resource section of this book has a link to a "do-it-yourself" audit produced by the U.S. Department of Energy. This resource allows individuals to make decisions about what investments would increase their home's overall energy efficiency.

Approximately 25% of home-energy use powers lights and appliances, so updating your appliances to those with higher energy-efficiency ratings could have a significant impact on your home's energy

Table 6. Steps that individuals can take in their home to reduce their energy consumption.

Energy Saving Activity	% Energy Saved per activity
Changing lightbulbs to CFLs	80% per bulb
Upgrading 1970s model appliances to Energy Star[114] appliances	50%
Unplugging appliances at night	Up to 5%
Insulating your water heater	Up to 40%
Replacing windows and insulating floors	Up to 40%
Turning thermostat below 68°F	3-5% per degree

use. Compact fluorescent lightbulbs (CFLs), for example, are 75-80% more efficient than incandescent lightbulbs. Table 6 shows a few easy actions that can lower energy use in the home, with the amount of energy that can be saved by each activity. Above the level of the individual, building codes can contribute significantly to energy efficiency of new homes.[115]

Transportation: Approximately 28% of all energy used in the U.S. is used for transportation, and most (17%) is used in cars and trucks.[116] The burning of a gallon of gasoline in an average truck or car puts 26 pounds of CO_2 into the atmosphere. Yet there are a number of changes that you can make that will lessen your car's total emissions of CO_2 to the atmosphere.

First and foremost: drive less. The average American drives approximately 16,000 miles per year in a car. With an average fuel efficiency of approximately 12 miles per gallon in city driving, this means annually putting an average of 24,500 pounds of carbon into the atmosphere.[117] Alternatives to driving alone in your car include carpooling, using mass or self-powered transit, telecommuting, and driving more efficiently. Mass transit makes a big difference in per capita energy consumption. New York State, for example, has the second-lowest average energy use in the U. S., mostly because a large proportion of its residents live in and around the New York City metropolitan region and use mass transit frequently.[118]

If you drive, drive the most fuel-efficient car that you can. Buying a new car that gets 10 more miles to the gallon than your old car reduces the annual amount of CO_2 emitted by approximately 2,300 pounds.[119]

Drive smarter. Tuning your car engine can increase fuel efficiency and reduce emissions by up to 50%.[120] A dirty air filter in a vehicle can reduce its efficiency by up to 10%, and under-inflated tires can reduce efficiency by 5% or more. Even ensuring that the car's gas cap is properly tightened can increase fuel efficiency. "Jackrabbit" driving (fast acceleration and quick braking) can increase fuel consumption by 40%, and only shortens the time of the drive by 4%. Slow and smooth acceleration and speed maintenance is much more efficient, and maintaining a lower cruising speed at 55 instead of 75 miles per hour can decrease fuel consumption by 20%. Idling wastes fuel, so if stopping for more than 30 seconds, turn your engine off. Avoid using air conditioning because it can decrease fuel efficiency by 10%, particularly in city driving. At high speeds, however, it is more efficient to use the air conditioner than open windows, which increases drag on the car.[121]

Above the level of the individual, there are many steps that can be taken to reduce CO_2 from cars and trucks. Car companies can produce more fuel-efficient fleets of vehicles. Increasing the average fuel efficiency of cars sold in the U.S. by just one mile per gallon could save nearly 1,500 pounds of carbon per car, per year, from being added to the atmosphere.[122] Companies will build what they think people want to buy, which means that individual purchasing decisions can make a difference. However, governments can also increase the requirements for new vehicle fuel efficiency. They can also invest in and/or encourage development of more mass transit systems. This can be something as simple as HOV (High Occupancy Vehicle) lanes on highways to encourage carpooling to construction of light-rail systems.

9.3 Energy Production

Approximately 80% of the energy produced in commercial power plants in the U.S. comes from burning oil, coal, or natural gas (Figure 44). Worldwide, fossil fuels provide 86% of powerplant energy. Coal, oil, and natural gas, however, are not equal in the amount of energy that they produce per pound of carbon emitted. One million BTUs[123] of energy produced from burning coal releases 205-227 pounds of carbon, whereas one million BTUs produced from oil releases 140-160 pounds and one million BTUs from natural gas burning releases 117 pounds of carbon.[124] Thus, the amount of energy produced from burning coal releases significantly more CO_2 into the atmosphere than producing the same amount of energy produced from burning natural gas. In fact, depending on many factors (such as the type of coal-fired power plant, extraction methods for both natural gas and coal from the ground, etc.), power plants fueled by natural gas emit up to 50% less CO_2 than power plants fired by coal.[125] Many other forms of en-

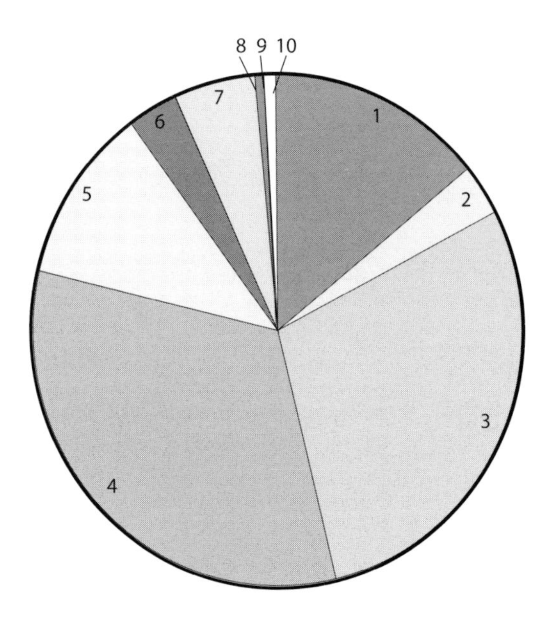

Figure 44. *Energy production in the United States. 1 – crude oil (14.3%); 2 – natural gas liquids (3.3%); 3 – dry natural gas (28.7%); 4 – coal (32.4%); 5 – nuclear electric power (11.5%); 6 – hydroelectric power (3.3%); 7 – biofuels (wood and waste, 5.3%); 8 – geothermal (0.5%); 9 – solar energy (0.1%); 10 – wind energy (0.7%).*[126]

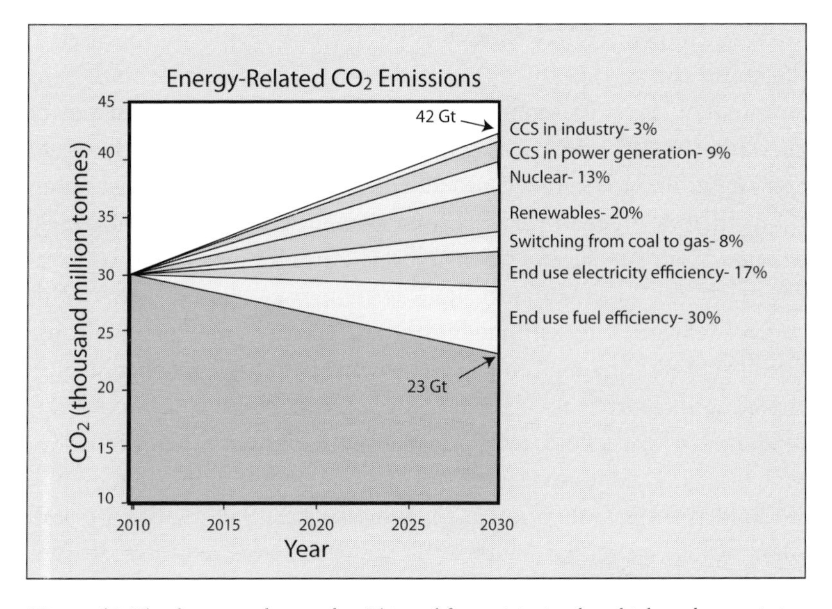

Figure 45. *The climate carbon wedge. This tool for envisioning how high-carbon-emitting energy sources can be replaced by "alternative" sources suggests that the solution lies not in just one alternative, but in harnessing the many possible energy sources together.*[127]

ergy generation – such as wind turbines, solar panels, and nuclear plants – produce little or no greenhouse gases. These are sometimes referred to as "alternative" or "renewable" energy sources.

Each of these energy-generating options – from coal to wind – comes with a multitude of different economic, political, and social costs and benefits. It can seem impossibly complex to even consider how to figure out which ones are the best for particular times or places. The **climate carbon wedge**, proposed by Princeton University professors Rob Solocow and Stephen Pacala, provides a tool for envisioning how high-carbon-emitting energy sources can be replaced gradually by "alternative" sources (Figure 45).[128] The wedge suggests that the solution lies not in just one alternative, but in harnessing the many possible energy sources together to fulfill the world's energy demands while reducing the amount of CO_2 released into the atmosphere. Like many issues discussed in this book, this is a vast and complex subject. Here we will only sketch briefly some of the costs and benefits of some of these alternative sources.

Wind power: Wind energy is the fastest growing source of electricity in the world, and is certainly one of the cleanest and most sustainable. In 2005, enough wind-power capacity was installed in the U.S. to power approximately 650,000 homes. Areas with a concentration of windmills are called windfarms. Concerns surrounding them include their impact on migratory bird and bat populations, aesthetics, and noise. Another obstacle is sometimes called "Not In My Back Yard" (NIMBY): people are frequently uncomfortable with unfamiliar things in their neighborhood. The costs of wind energy are currently relatively high, but government incentives are encouraging more manufacturers to enter the industry and develop new technologies, and so costs could decline in the not-too-distant future.[129]

Solar power: All of the world's existing energy reserves of coal, oil, and natural gas are matched by the global energy of just 20 days of sunshine across the world's surface. Of course, capturing and using even a fraction of this energy has proven to be extremely difficult. The main technology for collecting solar energy and turning it into a usable form of energy uses photovoltaic cells, which convert sunlight directly into electricity. A significant problem with current solar energy technology, however, is that it is difficult to store energy captured by solar panels for use during "down" times when it is cloudy or dark.[130]

Nuclear power: Nuclear power releases no greenhouse gases. In some countries, such as France, it provides a significant proportion of the electrical demand. In the U. S., however, it has long been controversial, mainly because of the problem of disposing of the dangerous radioactive waste that it generates. Even with the small amount of nuclear energy production currently in the U.S., from 1995 to 2008 nuclear energy production saved over 9,400 million tons of CO_2 from entering the atmosphere.[131]

Geothermal power: Geothermal energy is produced from the heat stored below the Earth's surface. One form of geothermal heating directly uses heat emanating from below Earth's crust. In most places, this requires drilling thousands of feet beneath the surface, but in some geologic settings (like Iceland) such heat can be tapped near the surface. Converting the energy into electricity, however, must be done in a power plant, and presently, because of technological limitations, this can only occur where there are heated reservoirs of water or steam in the Earth. Another form uses the relatively constant temperature several hundred meters beneath Earth's surface (approximately 10°C or 50°F). Geothermal energy can be used to heat or cool buildings anywhere on the planet; it generated approximately 4% of the renewable-energy based electricity consumed in the U.S. in 2007.[132] Domestic geothermal resources, however, could supply up to 30,000 years worth of energy to the U.S. at the current rate of national consumption, but it has not reached its full potential because of issues, including with technology, historically low natural gas prices, and public policies.[133]

Hydropower: Electricity generated from moving water is a clean, renewable, and historical form of energy production, and is the world's largest renewable energy resource. There are, however, environmental problems associated with building dams, such as sediment loss downstream and the disturbance of habitats for native species, such as salmon and freshwater mussels. Larger impacts, like flooding, dam failure, and even earthquakes, can also change regional hydrology.[134]

9.4. Carbon Sequestration

Carbon sequestration is the process of capturing CO_2 through biological or physical processes, and removing it from the atmosphere. For example, CO_2 can be captured and stored (sequestered) in trees through properly planted and managed forests. Of course, using vegetation to sequester CO_2 is only a temporary solution, because eventually the plants die, releasing the stored CO_2 back into the atmosphere. Maintaining healthy, sustainable forests is the only way to make this a long-term solution.

Carbon dioxide can also be captured during the process of burning fossil fuels. After being captured (*e.g.*, before it goes up the smokestack of a power plant), the CO_2 is then sequestered by being injected deep underground and stored in geologic traps. Large-scale use of this approach is still in the experimental stages, but technological advances could make this type of carbon sequestration economically viable in the next decade or so.[135]

Summary: What Can We Do?

The challenge of climate change could be the largest and most complex problem humanity has ever confronted. It is far bigger and harder to solve, for example, than any of the other massive collective technical projects – such as the dikes of The Netherlands, the Great Wall of China, the Great Pyramids, or manned space flight – which our species has previously undertaken. It requires mobilizing not just a few people in a few nations, but a large proportion of all the people on Earth. It requires going against the current of powerful economic, political, and social trends and interests. It requires much more than just personal action by individuals to conserve or recycle or use less energy; it requires large and difficult decisions and actions by local and national governments. Yet, in the end, it is only individuals who can compel their governments to act.

10. SOME FREQUENTLY ASKED QUESTIONS

(1) If geology tells us that the Earth's climate has changed in the past, why should we be concerned that it is changing now?

Over the past billion years, the Earth has experienced climates that were much warmer than those of today, as well as much colder periods. This fact, however, does not necessarily mean that future warming is nothing to worry about.

First, the conditions during those previous very hot or very cold times would hardly have been hospitable to modern humans. For example, between approximately 750 million and 580 million years ago, the Earth was in the grip of an ice age more extreme than any since, with most or all of the planet's surface covered with ice. At the height of the Age of Dinosaurs, 65-100 million years ago, global temperatures were as much as 9°C (16°F) higher than today, and sea level was perhaps 100 meters (328 feet) higher than today (so, you could have sailed a boat from the Gulf of Mexico to Alaska through Nebraska). Although scientists do not expect either of these extremes in the forseeable future, changes that *are* expected could cause enormous harm.

Second, it is not just warming, but also its rate that is the issue. The current rate of warming is very high compared to most climate changes of the geologic past, and relative to the ability of either humans or other organisms to readily migrate or adapt. Thus, although it is certainly true that life did not experience major extinctions with every climate oscillation in the geologic past (not even during the glacial-interglacial variations of the last 2.5 million years), most of these variations did not take place as rapidly as current changes are happening.

Third, and probably most importantly, humans have built a global civilization around the relatively stable climate conditions of the past 10,000 years. Even if species extinctions were not greatly increased, and even if humans in some developed countries were, on the whole, insulated from changes going on in their area, there are vast parts of the world where millions of people are much more sensitively connected to issues of their environment and cannot readily move or otherwise adapt in a short span of time.

❧ *See Chapters 1 and 4 for more on this topic.*

(2) How can we be sure that changes going on now are not just part of natural climate variation?

First, there are no known sources of natural variation that would give rise to changes as rapid as those observed in global temperature over the past 150 years. Second, there is a human-induced cause (increase CO_2) that not only fits the variation extremely well, but has long been expected to give rise to such change based on basic physical principles.

We have, of course, seen variations in climate over the past several thousand years, some of which are not well explained. Is it possible that current changes are just one of those fluctuations? The reason most scientists do not think so is that (a) the magnitude of the current trend is larger than any that we have seen in the past several thousand years, and (b) the directionality of the current trend shows an increasing rate of warming through time. Such variation does not look like any of the temperature changes seen in the past 10,000 years or so.

Could current change just be unusually extreme variation that we do not yet understand? Yes. In science (by definition), all phenomena are open to new explanations, so scientists must always be ready to consider other options. Scientists do not, however, favor or give equal weight to random or unknown variation if another known explanation fits the available data. Some explanations are better than others, and it would be ineffective to act as if every explanation, no matter how unlikely, should receive equal treatment.

See Chapters 3 and 4 for more on this topic.

(3) Why do most climate scientists believe that global warming is due to human activity rather than natural variation?

There are three major reasons why most climate scientists are convinced that the current warming is not due to natural processes:

(a) We know that greenhouse gases are accumulating in the atmosphere at levels that have not been experienced in over 20 million years.

(b) The pattern of the observed warming fits the pattern that we would expect from warming caused by the buildup of greenhouse gases. (That is, almost all areas of the planet are warming; the Earth's surface and lower atmosphere are warming; the upper atmosphere is cooling; the temperature changes are greatest in the Arctic during winter.)

(c) The warming is much more rapid than most of the natural variations we have seen in the past.

The past century of warming cannot be explained without factoring in anthropogenic influences. Only by including the net effect of human-made greenhouse gases and aerosols, can the observed changes be reproduced to match the actual record.

There are uncertainties in many widely accepted scientific theories, but this does not keep them from being widely accepted and used. Quantum Theory, for example, has many unresolved questions, yet it has been a very successful accomplishment responsible for the many electronic wonders of our age. Evolutionary Theory has uncertainties, yet forms the basis for most of biology.

See Chapters 5 and 6 for more on this topic.

(4) Correlation is not proof of causation. Isn't it true that there is simply no proof that CO_2 is the cause of the current warming?

There is no "proof" of anything in science. That is a property of mathematics. In science, one must look at the balance of observable evidence and formulate hypotheses that can explain this evidence. When possible, scientists make predictions and design experiments to confirm, modify, or contradict their hypotheses, and they modify their hypotheses as new information becomes available.

In the case of the hypothesis of anthropogenic global warming, we have an idea (first conceived over 100 years ago) that is based on well-established laws of physics, is consistent with extremely large quantities of observation and data, both contemporary and historical, and is supported by very sophisticated and refined global climate models that can successfully reproduce the climate's behavior over the last century. Given the lack of any extra planet Earths or a time machine, it is simply impossible to do any better than this.

See Chapter 2 and 6 for more on this topic.

(5) Globally averaged, 2008 was a cooler year than 2007. Does that mean that global warming isn't happening any more?

The year 2005 was the warmest year, globally, in recorded history, but 2008 was only the ninth warmest year on record.[129] However, this does not mean that the overall trend in global warming has been reversed. Indeed, this argument represents a fundamental misunderstanding of the difference between weather and climate. Climate is defined as the weather conditions averaged over a long period, usually decades. Thus, we cannot discern a trend in climate change by looking at a small number of years, much less a single one.

The last few years have remained far above a global baseline average temperature (such as the average of global temperatures from 1951 to 1990; see Figure 25). Looking at a graph of the directly recorded global temperatures over the past 150 years, we can see that even in globally and seasonally averaged and smoothed data, there are still numerous peaks and troughs that are irrelevant to the longer-term trends. The last four or five years do look as though the trend has paused or even reversed, but this is an artifact of how the graph was produced. Because the graph was made using a ten-year rolling average, it is actually too soon to detect the real trend direction. There is no convincing reason to think that the long-term warming trend has reversed, nor that it is likely to for many years to come.

❧ *See Chapter 4 for more on this topic.*

(6) Why would global warming necessarily be bad for humans?

It is true that many places with cold climates will benefit in some ways from global warming. These positive effects include longer growing seasons and greater agricultural productivity in high-latitude countries like Canada and Russia, smaller winter heating bills, and fewer hassles with icy roads. Some of the negative changes that will occur might not be especially difficult to deal with; for example, barring a precipitous disintegration of the continental ice sheets, the rise in sea level might be so gradual that in relatively wealthy cities like Seattle or New York, adjustments might be possible without a great deal of human hardship.

However, there will also almost certainly be a larger number of more significant negative consequences. For example, high latitude ecosystems – such as the Arctic and Antarctic – will change dramatically, leading to reduction or extinction of many species. And as the permafrost melts, buildings and roads built on it could sink, tilt, or collapse entirely. Droughts and severe weather events such as floods and tropical storms could become stronger and more frequent. Rising sea level will be difficult to deal with in densely-populated low-lying coastlines like those in Louisiana and Bangladesh and on coral atolls, and poorer countries will be disproportionately affected. Some human diseases could spread into areas in which they were formerly not a problem. Drinking and agricultural water might be harder to obtain. Ocean acidification will further endanger already-threatened coral reefs and other marine life.

Increasing temperatures also mean that climates (and the ecosystems associated with them) will shift poleward. Poisonous insects, snakes, and plants that thrive in warmer climates will be able to migrate as temperatures increase. Organisms that carry human diseases, like mosquitos, will have more reproductive cycles as temperatures increase. Farmers will have to grow different crops that are more suited to warmer climates, and will have to deal with pests that were previously not a problem to them.

❧ *See Chapter 7 for more on this topic.*

(7) Won't new technologies and "green energy" get us out of this?

They might. Humans might be able to invent and deploy technologies that produce or use energy that does not contribute to global warming. We might also be able to engineer effective ways to sequester the carbon that we are currently emitting and store it below ground. We can also much more widely employ some already available energy efficient technologies. It is important to recognize, however, that we are not even close to these technological solutions; they will take decades, at best, to make a significant difference.

Even if we were to stop fossil fuel burning altogether and immediately, temperatures would almost certainly continue to rise because of the additional CO_2 already in the atmosphere. Carbon dioxide concentrations would eventually start to decline, but it would take much longer for them to return to preindustrial levels than they did to build up. This means that even in this extreme "best case" scenario, we still need to expect some level of climate change and to prepare accordingly.

❧ *See Chapter 9 for more on this topic.*

(8) Aren't people who are arguing that global warming is going to happen just being alarmists? Will it really be that bad?

It really could be. This book discusses some of the possible serious impacts of global warming on humans, such as health, agriculture, land use, and water availability. These are not science fiction. There are steps that we can take to slow or cut emissions globally to avoid these scenarios, but our current lifestyle, including the ability to feed a large number of people, move vast quantities of goods large distances quickly, and live in the wide range of environments that we do, revolves around our climate system remaining relatively stable. Realistic worst-case scenarios of climate change could very plausibly lead to massive disruption of modern lifestyles, the global economy, and even national security.

❧ *See Chapter 9 for more on this topic.*

(9) Aren't there some scientists who argue that the climate isn't warming, or that if it is, humans aren't responsible?

Yes, there are scientists who question both that the climate is warming and that any such changes that are occurring are the result of human activity. The vast majority of climate scientists, however, agree that the planet's climate is warming, and that human activity is responsible. This is why we can speak of an overwhelming consensus in the scientific community about global warming and its causes.

❧ *See Chapter 8 for more on this topic.*

(10) We can't even forecast the local weather for next week with reasonable accuracy! How can we predict the climate over the next 100 years?

Although weather and climate are complex systems, that does not mean that they are entirely unpredictable. The unpredictable character of complex systems arises from their sensitivity to changes in the conditions that control their development. Weather is a highly complex mix of events that happen in a particular locality on any particular day; even small changes in rainfall, temperature, humidity, etc., can cause weather to vary wildly. Climate, however, is the broad generalization about a region's weather – the average over decades of weather patterns in a region. Although weather changes rapidly on human timescales, climate changes fairly slowly. Getting reasonably accurate predictions is a matter of choosing the right timescale – days in the case of weather, and decades to centuries in the case of climate.

The differences between predicting daily weather and predicting long-term climate patterns relate to the level of detail being modeled. As a rough analogy, consider a car riding down the highway creating turbulence in the air moving around it as the car moves, whirling the air in all directions like a tornado. That is analogous to weather, in that the exact air movement (direction and speed) at a specific point and time are very difficult to predict. If, however, you are driving at 60 miles per hour, and if the air is otherwise still, it is reasonable to predict that, overall, the average air speed passing the car will be 60 mph. If you accelerate to 70 mph, the air passing the car will still be turbulent, but it will have an average relative speed of 70 mph. The difference between the complex motion of the air around the car and the movement of the car through the air is something like the difference between weather and climate. Another analogy involves the game of pinball: you cannot predict the exact path that a ball will take as it bounces through a pinball machine, but you can predict that the ball will eventually land in the lowest part of the machine and end your game.

Thus, although we cannot predict the weather in a particular place and on a particular day (or year) in 100 years time, we can predict that on average it will be far warmer if the concentration of greenhouse gases in the atmosphere continues to rise.

See Chapter 6 for more on this topic.

(11) A lot of climate predictions depend on computer models. How much can we trust such models?

Climate is an average of weather over decades. It can vary seemingly unpredictably, but can truly only move within the limits set by major influences like the Sun and levels of greenhouse gases in the atmosphere. We can be confident, for example, that the summers will be warmer than winters for as long as the Earth's axis remains tilted, even if we cannot guess the exact temperatures of specific seasons (a weather prediction). In this sense, certain influences yield, within a given margin of error or uncertainty, certain results. This is why models, which are based on such cause-and event relationships, allow us to predict the future with some confidence.

Furthermore, the validity of models can be tested against climate history. If they can predict the past (which the best models do pretty well), they are probably on the right track for predicting the future. Moreover, models designed two decades ago are not far off in their predictions of the climate that we are currently experiencing.

Critics of models have valid points about degrees of confidence in a single model, but this does not mean, as is sometimes challenged, that the models must be biased toward alarmism, that is, toward greater climate change. It is just as likely that these models err on the side of caution. Most modelers accept that despite constant improvements over more than half a century, there are uncertainties in global climate models. They acknowledge, for instance, that one of the greatest uncertainties in their models is how clouds will respond to climate change. Their predictions usually come with generous error bars.

☙ *See Chapter 6 for more on this topic.*

(12) Couldn't the Sun be responsible for the observed recent climate changes?

No one denies our star's central role in determining how warm our planet is. Switch off the Sun and Earth would become a very chilly place. The issue today is how much solar changes have contributed to the recent warming, and what that tells us about future climate. The current scientific consensus is that changes in the energy output from the Sun do not successfully account for the current warming trend. Eleven-year sunspot cycles have been consistently recorded, but these have risen and fallen as expected, never increasing their net output of energy. No other solar outputs correlate with the warming trend, either. So for the period for which we have direct, observable records, the Earth has warmed dramatically even though there has been no corresponding rise in any kind of solar activity.

☙ *See Chapter 2 for more on this topic.*

(13) Didn't scientists predict global cooling in the 1970s?

Yes. A handful of scientific papers discussed the possibility of a new ice age at some point in the relatively near future, leading to some pretty sensational media coverage. One of the sources of this idea might have been a 1971 presentation by Stephen Schneider, then a climate researcher at NASA's Goddard Space Flight Center.[136] Schneider's highly visible talk, at the annual meeting of the American Association for the Advancement of Science, suggested that the cooling effect of dirty air could outweigh the warming effect of CO_2, potentially leading to an ice age if aerosol pollution quadrupled. This scenario was seen as plausible by many other scientists, in part because at the time the planet had been showing slight cooling. Much of the fuss was prompted by the media hype about the 1976-1977 and 1977-1978 winters, which produced some remarkably cold weather over the United States. Schneider soon realized, however, that he had overestimated the cooling effect of aerosol pollution and underestimated the effect of CO_2, meaning that warming was more likely than cooling in the long term. Another paper – a 1975 report by the U. S. National Academy of Sciences – merely called for more research to test the cooling hypothesis.[137]

In contrast, the calls for action to prevent further human-induced global warming are based on an enormous body of research by thousands of scientists over more than a century that has been subjected to intense – and sometimes ferocious – scrutiny. Surveys of the scientific literature have found that between 1965 and 1979 (before attention to "global warming" began accelerating in the 1980s), 44 scientific publications predicted warming, 20 were neutral, and only 7 predicted cooling.[138] So although predictions of cooling got more media attention, the majority of scientists were predicting warming even then. Now, according to the latest IPCC report,[139] it is more than 90% certain that the world is already warming as a result of human activity.

Thus, the cooling "scare" involved a handful of scientists and lasted only a few years, until new evidence caused it to be rejected. In contrast, nearly the entire scientific community is now concerned about global warming and this concern has been steadily growing as the evidence accumulates.

See Chapter 6 for more on this topic.

(14) Aren't human CO_2 emissions too tiny to matter?

Ice cores show that carbon dioxide levels in the atmosphere have remained between 180 and 300 parts per million for at least the past 800,000 years. In recent centuries, however, CO_2 levels have risen to at least 380 ppm. It is true that human emissions of CO_2 are small compared with natural sources. However, the fact that CO_2 levels have remained steady until very recently shows that natural emission sources are usually balanced by natural absorption sinks. For the past 150 years or so, slightly more CO_2 is entering the atmosphere than can be soaked up by carbon sinks.

The consumption of rotting vegetation by animals and microbes emits approximately 220 gigatons of CO_2 every year, and respiration by living plants emits another 220 gigatons (Gt). These huge amounts are balanced by the 440 Gt of CO_2 absorbed from the atmosphere each year as land plants photosynthesize. Similarly, parts of the oceans release approximately 330 Gt of CO_2 per year, depending on temperature and rates of photosynthesis by phytoplankton, but other parts usually soak up just as much – and are now soaking up even more. Human emissions of CO_2 are now estimated to be 26.4 Gt per year, up from 23.5 Gt in the 1990s, according to the 2007

IPCC report.[140] Disturbances to the land – through deforestation and agriculture, for instance – also contribute roughly 5.9 Gt per year. Approximately 40% of the extra CO_2 entering the atmosphere due to human activity is being absorbed by natural carbon sinks, mostly by the oceans. The rest is boosting levels of CO_2 in the atmosphere.

Claims that volcanoes currently emit more CO_2 than human activities are simply not true. In the very distant past, there have been intervals of much more active volcanism than at present, and also single volcanic eruptions so massive that they covered vast areas in lava or ash and appear to have released enough CO_2 to warm the planet (after the initial cooling caused by the dust). Measurements of CO_2 levels over the past 50 years, however, show no significant rises after single eruptions. Total present emissions from volcanoes on land, for example, are estimated to average just 0.3 Gt of CO_2 each year – approximately 1% of human emissions.

☙ *See Chapter 5 for more on this topic.*

(15) Why does CO_2 matter so much if it isn't the most important greenhouse gas?

Water is, molecule-for-molecule, a more powerful greenhouse gas than CO_2. Excess water vapor rains out in days. So, does water or CO_2 contribute more to climate change? It is not surprising that there is a lot of confusion about this – the answer is far from simple.

First, there is the greenhouse effect, and then there is climate change. The greenhouse effect currently keeps our planet 20° to 30°C warmer than it would be otherwise, allowing life on our planet to thrive. Global warming is the rise in temperature caused by an increase in the levels of greenhouse gases. Water vapor is by far the most important contributor to the greenhouse effect. Approximately 50% of the greenhouse effect is due to water vapor, with another 25% due to clouds, 20% due to CO_2, and with other gases accounting for the remainder.

So why are climate scientists not a lot more worried about water vapor than about CO_2? The answer has to do with how long greenhouse gases persist in the atmosphere. Excess CO_2 accumulates, warming the atmosphere, which raises water vapor levels and causes further warming. This rapid turnover of water vapor means that even if human activity was directly adding or removing significant amounts of water vapor (which it is not), there would be no slow build-up of water vapor as is happening with CO_2. The level of water vapor in the atmosphere is determined mainly by temperature, and any excess is rapidly lost. The level of CO_2 is determined by the balance between sources and sinks, and it would take hundreds of years for it to return to pre-industrial levels even if all emissions ceased tomorrow. To put this another way, there is no limit to how much rain can fall, but there is a limit to how much extra CO_2 that the oceans and other sinks can soak up.

Of course, CO_2 is not the only greenhouse gas emitted by humans. Many greenhouse gases, such as methane, are far more powerful than CO_2 in terms of infrared absorption per molecule. However, the overall quantities of these other gases are tiny. Even allowing for the relative strength of the effects, CO_2 is still responsible for 60% of the additional warming caused by all the greenhouse gases emitted as a result of human activity.

❧ *See Chapter 2 for more on this topic.*

(16) Some ice cores show that CO_2 increases lag behind temperature rises. Doesn't this disprove the link to global warming?

Ice cores have been used to correlate CO_2 and temperature over the past 800,000 years. In these cases, CO_2 increases and decreases can lag behind temperature increases and decreases. It takes approximately 5,000 years for an ice age to end and, after the initial 800-year lag, temperature and CO_2 concentrations in the atmosphere rise together for another 4,200 years. The lag indicates that rising CO_2 did not cause the initial warming as past ice ages ended, but it does not in any way contradict the idea that higher CO_2 levels cause warming. Sometimes a house gets warmer even when the central heating is turned off; however, this does not prove that the central heating does not work – it might have been a hot day outside, or maybe the oven was left on. Just as there is more than one way to heat a house, so there is more than one way to heat a planet. What happened between glacial and interglacial times to trigger the initial warming was caused by another factor: changes in the shape of Earth's orbit that caused changes in the amount of energy from the Sun absorbed by Earth, that in turn caused a rise in temperature. This led to the release of CO_2 from the ocean, because it is less soluble in warmer water. The shrinking of the ice sheets further amplified the warming because of negative albedo. The increased CO_2 then amplified these changes by increasing temperature further – this is part of a positive feedback cycle. The initial warming also melted regions of permafrost, which are also carbon sinks. Melting led to a release of CO_2 that led to temperature increase, which became part of another positive feedback cycle contributing to the warming of the planet.

In summary, the lag in CO_2 seen in ice core data only shows that increases in atmospheric CO_2 levels were not responsible for the initial temperature increase. However, the initial temperature change, caused by an increase in the amount of energy absorbed from the Sun due to variations in Earth's orbit, caused the release of CO_2 from carbon sinks in the ocean and permafrost. This triggered a positive feedback cycle that warmed the Earth.

☙ *See Chapters 4 and 6 for more on this topic.*

(17) Hasn't the "hockey stick" graph of temperature been proven wrong?

Numerous studies support the key conclusion of the classic "hockey stick" graph: Earth's temperature is warmer than it has been in at least 1,000 years (see Figure 24). The graph was the result of the first comprehensive attempt to reconstruct the global average temperature over the past 1,000 years, based on numerous proxies. It shows that global average temperature was holding fairly steady until the last part of the 20th century and then it suddenly shot up. This correlates very well with CO_2 data from ice cores for the last 1,000 years (see Figure 23). Based upon temperature and CO_2 reconstructions for the last 800,000 years, scientists have shown that CO_2 and global temperature co-vary, and because CO_2 levels have increased dramatically over the past 2,000 years, global average temperature is expected to continue to rise dramatically. The conclusion that humans are making the world warmer does not depend on reconstructions of temperature prior to direct records.

The "hockey stick" graph first appeared in a paper published in 1998, and the graph was highlighted in the 2001 IPCC report.[141] It is true that there are uncertainties about the accuracy of all past reconstructions, and that these uncertainties have sometimes been ignored or glossed over by those who have presented the "hockey stick" as evidence for human-caused global warming. Yet since 2001, there have been repeated claims that the reconstruction is, at best, seriously flawed and, at worst, a fraudulent result due to an artifact of the statistical methods used to create it. Details of the claims and counterclaims involve lengthy statistical arguments, but a good summary was provided in the 2006 report of the U.S. National Academy of Sciences. The academy was asked by Congress to assess the validity of temperature and CO_2 reconstructions, including the hockey stick. The report stated that "the late 20th century warmth in the Northern Hemisphere was unprecedented during at least the last 1000 years."[142] This conclusion has subsequently been supported by an array of evidence that includes both additional large-scale surface CO_2 and temperature reconstructions and pronounced changes in a variety of local proxy indicators, such as melting ice caps and the retreat of glaciers around the world. Most researchers agree that although the original "hockey stick" graph can – and has – been improved in a number of ways, it was not far off the mark. Most subsequent temperature reconstructions fall within the error bars of the original graph. Some show far more variability leading up to the 20th century than the "hockey stick," but none suggest that global average temperature has, at any time in the past 1,000 years, been higher than in the last part of the 20th century.

❧ *See Chapters 4 and 5 for more on this topic.*

(18) Aren't there all kinds of objections to the IPCC?

The Intergovernmental Panel on Climate Change (IPCC) was set up by the United Nations in 1988 to review scientific conclusions on climate change that had already passed peer review and been published. It includes thousands of climate scientists and is open to all member countries of the United Nations and the World Meteorological Organization. The IPCC has published four reports so far – in 1990, 1995, 2001, and 2007. The 2007 report[143] concluded that large-scale recent warmth likely exceeds the range seen in past 1,300 years, and that it is "very likely" (> 90% probability) that humans are responsible. The remaining uncertainties in the latest report mainly involve the precise nature of the changes to be expected, particularly with respect to sea-level rise, El Niño changes, and regional hydrologic change – drought frequency and snow pack melt, mid-latitude storms, and hurricanes. These uncertainties do not, however, diminish the credibility of the far more solid conclusions above. This report has nevertheless been criticized by many people on a variety of grounds.

The process of producing the 2007 IPCC report entitled "Summary for Policy Makers" (SPM) – the piece written for nonscientists – was also criticized as having been the result of negotiation among bureaucrats instead of solid science. This was not what happened. Government representatives from all participating nations took the draft summary (written by the lead authors of the individual chapters) and discussed whether the text truly reflected the underlying science in the main report. It is key to note that the text of the leading authors, who were scientists, was edited on behalf of the governments for whom the report was being written so that conclusions could be understood by their (mostly nonscientific) officials. It is also important to note that the scientists had to be content that the final language that was agreed upon conformed with the underlying science in the technical chapters. The advantage of this process was that everyone involved was absolutely clear about what was meant by each sentence. This process also was supposed to allow the governments involved to feel as though they "owned" part of the report, and gave the governments a vested interest in making the report as good as it could be, given the scientific uncertainties. There were in fact plenty of safeguards (not the least the scientists present) to ensure that the report was not slanted in any one direction.

See Chapter 8 for more on this topic.

(19) Is climate change too big or too far along to be stopped?

It might be. All human societies have developed with anticipation of certain climate patterns and have always depended on climate-dependent natural resources. In the Pacific Northwest, for example, people rely on winter snow pack to store water that arrives in the winter for delivery in the summer when their demands are highest. Because significant climate change would alter accustomed climate patterns and regional natural resources (some natural systems will be irreversibly damaged by global warming), it could pose disruptions to socioeconomic systems around the world. These disruptions would be worse where global warming worsens existing conflicts over scarce resources and where the funding or capacity for preparing for or adapting to these changes is lacking. Many people agree that global warming is likely to have worse consequences for those with the least resources and therefore least able to adapt – the economically or politically vulnerable, for example.

"Stopping" anthropogenic global warming completely is now widely viewed as impossible in the short-term.[144] There is, however, still time to minimize it, and there are things that we can do. Global warming *is* reversible in the long-term, at least in the sense that we could, eventually, bring global greenhouse gas emissions, the "human" part of climate change, back down. The issue here is a matter of how long we can wait. The sooner we reduce emissions, the better.

❧ *See Chapters 6 and 7 for more on this topic.*

(20) What can be done by the average citizen?

We can take *individual* actions; for example, we can reduce actions that rely on fossil fuels, *e.g.,* in transportation, home heating and lighting, and consumption of goods. And we can take *collective* actions, such as encouraging friends and neighbors to do the same, elect politicians who will make the needed policy changes, and work for broad societal change.

Shifting the world's economy away from its dependence on fossil fuels is the single step that would do the most to reduce anthropogenic climate change. This is also a huge challenge that will not be accomplished by any one change. It will require actions big and small by individuals, corporations, and governments around the world. Although people sometimes feel that nothing that they do matters, the only thing that *does* matter is what people do.

Because global warming is caused by emissions of greenhouse gases, anything that a person does that reduces those emissions will help to counteract global warming. In our personal lives, driving less, carpooling, and switching to vehicles that get more miles to the gallon are all ways that we can reduce emissions.

Buying food that is in season and has been grown locally reduces the emissions that result from transporting food long distances. Decreasing energy use at home can help too. Beyond changing our own consumption and energy-use habits, we could encourage others to do so too. Transportation, energy use, and other policies at the local, state, and federal levels will all influence how much greenhouse gases are emitted. Some corporations are beginning to act to "offset" (reduce) their contributions to global warming. This is voluntary in the U.S., but mandatory in Europe. An individual might also think about the way a corporation is addressing global warming before making investment choices.

❧ *See Chapter 9 for more on this topic.*

(21) Can't we just figure out how to absorb CO_2 from the air as an alternative to reducing emissions from human activity?

Two ways of sequestering CO_2 are to (a) incorporate it into vegetation by planting forests, and (b) store it deeply underground. Planting forests can take up only a limited amount of CO_2, and this is only a temporary solution, because the carbon in logs will return to the atmosphere when the logs decay or burn. Deep burial is much more expensive, but it is currently the only long-term solution. Many scientists are currently working to improve deep-burial methods so that carbon sequestration can be implemented globally, but no solution has yet been found that is cost-effective and known to trap CO_2 effectively indefinitely. This is only part of the equation, anyway. We also need to reduce our emissions as a species, so that as our global population goes up, we can all live more efficiently and use carbon sequestration as one part of that solution.

❧ *See Chapter 9 for more on this topic.*

Figure 46. *Polar bear on an ice flow in Ukkusiksalik National Park, Nunavut, Canada. Polar bears have become "poster animals" for global climate change advocates, because their habitat on rapidly shrinking Arctic sea ice is decreasing at an alarming rate. Photograph by Ansgar Walk via Wikimedia Commons.*

Figure 47. *Teaching the teachers is the first step! 4-H Club Educators learn a simple classroom demonstration on rain formation at the Museum of the Earth in Ithaca, New York. Photograph by Paleontological Research Institution.*

11. TEACHING ABOUT CLIMATE CHANGE

Teachers, especially at the pre-college level, play a very important role in educating the public about climate change. As was discussed throughout this book, climate change is a very complex issue, and one must be familiar with a variety of sciences to truly understand the breadth of climate change. The nature of science itself is one of the most important subjects for helping students understand the study of climate change. As discussed in Chapter 8, the public conception of scientific certainty is different from that perceived by scientists.

Society is in the middle of a political debate about climate change, yet opinions are formed and decisions made by many without understanding the science behind proposed policies. Considering the forecasted future ramifications of climate change, climate change policy-making has been slow to develop and is still in its early stages. It is therefore important to teach the scientific principles behind climate and anthropogenic climate change to today's youth, so that their actions as citizens and voters will yield future policy that is informed by science.

Following are some suggestions for teachers about how to incorporate climate change into the classroom. Be sure to check the "Sources of More Information" section of this book for resources especially highlighted for teachers.

(1) Engage students' existing conceptions about climate and climate change.

Effective instruction requires that you engage what students already know and understand related to the topic you are teaching. They will already have ideas about climate, weather, and climate change, and regardless of their age or the setting, these existing conceptions will vary widely. Although the temptation is great to try and replace their naïve conceptions with scientific ones, learning is not built by swapping one conception for another. Existing conceptions almost always contain at least some aspects that are scientifically accurate. Drawing these out while diplomatically bringing evidence to bear on problems of the students' conceptions that are not scientific allows learners to use existing knowledge as building blocks for scientific understanding.

(2) Teach science in a science class, without advocating political positions.

Although climate change can be an emotional and political issue, it is important to separate feelings and convictions from scientific pursuits. Arrange your curriculum or lesson plans to anticipate, engage, and address their existing conceptions about human-made climate change, but do not assume that your students will necessarily know enough to have strong positions. In other words, be positive and do not start defensively. Resources to help you teach climate change abound, and you can find lists of some of them in the back of this book.

(3) Teach across the curriculum.

This might seem to slightly conflict with the previous suggestion, but there is a distinction between what is done in teaching the science of climate change and contextualizing the science of climate change across the curriculum. Climate is a complex and dynamic system. To understand the effects of climate change, we must learn aspects of science, mathematics, social studies, history, and communication. Basic skills such as addition, graph reading, and critical thinking are also integral to understanding climate change. Because mitigating climate change is interdisciplinary, teachers of social studies, English language arts, and others could collaborate through integrating climate change into their class at approximately the same time, so that a cohesive picture of climate change can form. (Of course, elementary teachers typically teach all of these.) An understanding of climate science will be enriched and made more durable if students are connected to the societal and cultural implications of changing climate. Although that principle should not be ignored entirely in upper-level science classrooms, it is primarily a discussion for social studies and humanities courses.

(4) Integrate climate change into the curriculum at multiple points throughout the year.

Erosion, weathering, flora, and fauna all vary with climate. The Sun is the engine that drives climate. Fossil fuels influence climate. The renewable energy sources of wind and solar energy are dependent upon climate factors. Because it is so multifaceted, climate change can be incorporated into lesson plans throughout the school year. Students can make their own temperature graphs and compare them to regional and global data. They can interpret scientific literature on climate change and report on it, or they can compare media coverage of various aspects of climate change. They can also participate in citizen science projects that enhance their knowledge of weather and climate. In some projects, students can participate as a class, whereas in others, they can individually create an inquiry-based project. In summary, it is important to encourage your students to think about this issue frequently, and to see it as part of a complex scientific understanding of our Earth, rather than as an isolated lesson.

(5) Don't be emotional or pessimistic.

One reason that climate change is so difficult to teach is because possible outcomes can be disheartening or even scary. Not only do polar bears feel negative consequences, but populated islands could become flooded and uninhabitable in other parts of the world. These are certainly sobering, emotional thoughts. And unless you are teaching social studies, explaining only the merits of various climate change mitigation strategies put forth by governmental institutions teaches little of the science behind the decision-making associated with why mitigation is important.

It is important, instead, to focus on what can be accomplished with respect to sustainability and mitigating climate change effects. Stick to positive reinforcement, and use examples of how being educated on issues like these has helped effectuate change in the past (*e.g.*, decreasing acid rain or saving certain endangered species). Local zoos and aquaria can be tremendous resources for providing examples of species and ecosystems that are being preserved through thoughtful action.

(6) Highlight opportunities for careers in alternative energies and energy efficiency, and engage students with projects in climate science and sustainability.

Alternative energy sources are popping up as new technologies are developed. Recruit representatives from wind, solar, energy conservation, or biofuel industries to participate in career days. Take advantage of curricula developed to teach the science behind alternative energy sources as a means to engage your students in climate science and sustainability. Talk to your local energy provider, who will most likely have resources available to you and can give a local perspective on a global issue.

(7) Get back to the basics – climate is a long-term, *global* trend.

It is difficult for some to grasp the long-term and global effects of climate change. Climate is the long-term weather pattern in an area, and individual years mean little without examining them in the context of the last 10 or 20 years. Also, climate change does not only, or always, equate to significant warming in your area. Polar regions experience much more warming than do temperate regions, and the overall global average temperature is predicted to rise only a few degrees. These small changes over long periods of time can trigger significant changes in global climate patterns. A globally averaged temperature decrease of only 7°C (12°F) globally was enough to create a mile-thick sheet of ice over the northeastern United States by the peak of the Last Glacial Maximum! Although most regions are expected to experience at least some temperature increase within the next century, for at least the next few decades, certain regions will experience almost no increase or even slight cooling in their climate regime; many areas will feel changing patterns in precipitation and wind rather more than temperature. Climate change is not just a temperature increase, and it will be felt differently everywhere.

(8) Use local, tangible examples of climate change, or global impacts that everyone will be familiar with.

Climate change is not causing the same effects in every place, but there are tangible expressions of the impacts of climate change on our environment, no matter your location. Using citizen science projects is a good way to begin searching for changes in your area that can be seen and measured in a classroom setting. The U.S. Geological Survey has put together climate change impacts by region throughout the United States.[145] For example, climate change might lower the level of Lake Erie, changing the habitable landscape for organisms and costing millions of dollars in infrastructure changes in ports and shorelines. Warmer Great Lake temperatures are increasing the amount of lake-effect snow. See the resources section in this book for further suggestions of how to incorporate local, tangible impacts of climate change in your community.

A number of species impacted by climate change provide excellent opportunities to engage students in considering the relationship between climate change and conservation. Many states could lose their official state birds if they head for cooler climates – including the Baltimore Oriole of Maryland, the Black-capped Chickadee of Massachusetts, or the American Goldfinch of Iowa. Polar Bears were put on the "threatened" list because of impacts on their environment from global climate change. The Pika in the western United States is currently under deliberation for being listed as "threatened" as well, and as of 2008, the Mississippi Gopher Frog had a population of over 100 individuals. In 1999, the extinction of the Golden Toad in Central America marked the first documented species extinction driven by climate change (see Figure 38). For more examples, see the references section of this book.

(9) Use examples from popular media to stimulate discussion and interest.

Science fiction writings and films, for good or ill, have impacted popular conception of many climate issues. Whether or not films like *The Day after Tomorrow* (Twentieth Century Fox, 2004) represent good science, they can serve as an engaging starting point for climate change discussions and projects. Although political, Al Gore's *An Inconvenient Truth* (film and book[146]) might also serve to stimulate discussion and yield good science project suggestions or classroom debates.

(10) Practice sustainability in the classroom.

It is important, after teaching about global climate change, to remain positive about how education and societal efforts can mitigate the effects of climate change. So it is crucial to practice sustainability in the classroom. Hang reminder signs around the classroom with sustainability tips. Even place them in student and teacher restrooms for high visibility. Turn lights out and unplug computers and other appliances when not in use. Encourage the cafeteria to purchase local, in-season fruits and vegetables instead of trucking them in from areas around the country. Use paper only when necessary, and be sure to use both sides of it! Use real cutlery instead of plastic, throw-away forks and plates at classroom parties. It takes a lot of energy to create the plastic forks and knives that we use once and then discard. Calculate your classroom's carbon footprint and aim to lower it by the end of the school year. All of these activities educate students about climate change and sustainable lifestyles that curb CO_2 emissions, and they empower students to make a difference.

(11) Use your local resources.

There are many institutions and individuals in most communities that have expertise on climate change. No doubt you use some of them already for career days and field-trip opportunities. These include natural history museums, zoos, aquaria, science centers, and local colleges. Natural history museums and science centers are frequently overlooked as sources of climate change information, but many have expertise in the area and have educators available to visit your classroom or make a presentation in their museum. Zoos and aquaria have an excellent vantage point on the biological impacts of climate change, because they are repositories of animals whose habitats have been altered by human activity, and their educators can discuss animals that are threatened or endangered due to climate change. In these cases, an activity worksheet during a field trip can easily incorporate aspects of climate change. Finally, many college departments have discipline-related student groups who participate in outreach activities. Geology, biology, and physics departments would be great places to begin searching for student expert classroom volunteers!

SOURCES OF MORE INFORMATION

Books

Archer, David. 2007. *Global Warming. Understanding the Forecast.* Blackwell Publishing, Malden, Massachusetts, 194 pp. An introduction for the general reader to all of the interconnected variables of climate and how they might react in the face of a changing climate.

Broecker, Wallace S., & Robert Kunzig. 2008. *Fixing Climate. What Past Climate Changes Reveal About the Current Threat – and How to Counter It.* Hill & Wang, New York, 253 pp. Two noted climate scientists give their perspective on climate and mitigation strategies.

Cronin, Thomas M. 2009. *Paleoclimates: Understanding Climate Change Past and Present.* Columbia University Press, New York, 448 pp. An excellent and thorough scientific discussion of paleoclimatology and what it has taught us about present and future climates.

Emanuel, Kerry. 2007. *What We Know About Climate Change.* Boston Review Books/ MIT Press, Cambridge, Massachusetts, 85 pp. A brief, clear, and easy-to-read essay by a distinguished climate scientist about how the climate works and how humans are affecting it.

Houghton, John. 2009. *Global Warming. The Complete Briefing, 4th ed.* Cambridge University Press, New York, 456 pp. An excellent, thorough review of the scientific information supporting climate change and possible outcomes of that change. Technical, but still accessible.

Intergovernmental Panel on Climate Change (IPCC). 2007. *Climate Change 2007 – The Physical Science Basis.* Cambridge University Press, New York, 996 pp. (also available at http://www.ipcc.ch). The authoritative scientific review of the scientific climate literature that contributed to our current standing on anthropogenic climate change.

Kump, Lee R., James F. Kasting, & Robert G. Crane. 2004. *The Earth System, 2nd ed.* Pearson Prentice Hall, Upper Saddle River, New Jersey, 420 pp. A college-level text in Earth system sciences, especially strong on the chemistry of the Earth's component systems.

Maslin, Mark. 2004. *Global Warming. A Very Short Introduction.* Oxford University Press, Oxford, U.K., 162 pp. An excellent, brief introduction to the issue for the general reader.

Mooney, Chris. 2007. *Storm World: Hurricanes, Politics, and the Battle Over Global Warming.* Houghton Mifflin Harcourt, New York, 400 pp. Semipopular but well-researched exploration of the science behind severe weather and its connection to climate change. Also discusses the role of the media.

Ruddiman, William F. 2008. *Earth's Climate. Past and Future, 2nd ed.* W. H. Freeman and Company, New York, 388 pp. An excellent, up-to-date, college-level climate science textbook with a very comprehensive explanation of the different facets of Earth's climate throughout geologic time.

Books for Younger Readers

Cherry, Lynne, & Gary Braasch. 2008. *How We Know What We Know About Our Changing Climate: Scientists and Kids Explore Global Warming.* Dawn Publications, Nevada City, California, 66 pp. An interesting explanation of the many lines of evidence for human-made climate change using examples of how youth in citizen science projects can make contributions to science.

David, Laurie, & Cabria Gordon. 2007. *The Down-To-Earth Guide to Global Warming.* Orchard Books, New York, 112 pp. A simplified account of the major concerns regarding global warming, and their potential solutions, using kid-friendly pop-culture references and engaging visual aids to connect with young readers. This book highlights sustainable or "green" jobs and ways that youth can become activists in their own communities.

Rockwell, Anne. 2006. *Why Are the Ice Caps Melting? The Dangers of Global Warming.* HarperCollins Publishing, New York, 33 pp. A simple scientific explanation of global warming and the potential impacts on the Earth. Strategies for young readers to contribute to the health of the planet are also included.

Sources Especially for Teachers

Books & Articles

Malnor, Carol L. 2008. *A Teacher's Guide to How We Know What We Know About Our Changing Climate: Lessons, Resources, and Guidelines about Global Warming.* Dawn Publications, Nevada City, California. 56 pp. An excellent teacher companion to Cherry & Braasch's book (listed under Books for Younger Readers), with activities and resources aligned to national standards and best practices. The activities are thoughtfully organized with easy-to-understand directions and useful worksheets.

Schmidt, Gavin, Joshua Wolfe, & Jeffrey D. Sachs. 2009. *Climate Change: Picturing the Science.* W. W. Norton and Company, New York, 320 pp. Scientific discussions coupled with moving and appropriate photography.

Shepardson, D. P., D. Niyogi, S. Choi, & U. Charusombat. 2009. Seventh grade students' conceptions of global warming and climate change. *Environmental Education Research*, 15(5): 549-570.

Websites

The Climate Literacy Network provides a wealth of information on many topics associated with climate change, including a compilation of resources specifically for teachers and students; http://climateliteracynow.org.

The Department of Ecology for the State of Washington has outreach materials on climate science, including teacher resources, implications of future climate change, and mitigation efforts in the region; http://www.ecy.wa.gov/climatechange/resources.htm.

ClimateChangeEducation.org is a virtual warehouse of well-recommended resources on climate change education. Sections of the website are devoted to lesson plans, class topics, and student project ideas; http://www.climatechangeeducation.org/k-12/index.html.

The National Center for Atmospheric Research provides a fun platform for teacher and student resources, including citizen science initiatives. It also provides material in English and Spanish; http://www.windows.ucar.edu.

The National Oceanic and Atmospheric Administration provides an excellent platform for teacher resources on anthropogenic climate change, paleoclimate research, and training opportunities for teachers; http://www.education.noaa.gov/teachers1.html.

Rutgers University Climate and Environmental Change Initiative provides classroom activities, educational resources and northeastern regional climate information, and opportunities for students to interact with real scientists in the region; http://climatechange.rutgers.edu/resources/education.php.

The American Association for the Advancement of Science provides a large number of references on climate change, including carbon footprint calculators and video material; http://www.aaas.org/news/press_room/climate_change.

The United States Global Change Research Program provides regional climate information, teachers guides for formal and informal educators, and other materials; http://www.globalchange.gov/resources/educators/toolkit/materials.

Citizen Science Opportunities

Journey North allows you to map "spring's journey north" using sunlight, plants, and animals with programs and resources provided for teachers, classrooms, and students. This information is important to scientists who, as climate warms and spring comes earlier, try to record how different organisms react to these changes; http://www.learner.org/jnorth.

Globe includes a consortium of citizen science projects on cloud cover, stars, climate, and the carbon cycle, written with teachers and their students in mind; http://www.globe.gov.

Project Budburst allows you to monitor many of your favorite flower types for the dates of budding and blooming to track the impact that climate change is having on flowering plants; http://www.windows.ucar.edu/citizen_science/budburst.

Monarch Watch allows you to monitor the travel of monarch butterflies across the continent so that scientists can learn how they are impacted by climate change; http://monarchwatch.org.

Tracking Climate in Your Backyard is a collaboration project between 4-H, the Paleontological Research Institution, and the citizen science precipitation monitoring network CoCoRaHS. Provided are a curriculum for formal and informal educators and the opportunity to monitor precipitation with other citizen scientists across the U.S.; http://www.museumoftheearthorg/TrackingClimate.

Project Feederwatch, BirdSleuth, Urban Bird Studies, and Nest Watch are all citizen science projects sponsored by Cornell University's Lab of Ornithology. Each project has different logistical requirements, and so can be accommodated by different classrooms with different needs; http://www.birds.cornell.edu.

Websites

http://www.realclimate.org is an excellent compendium of documents explaining the many different aspects of climate science. Although some references are scientific papers, many links to public opinion polls and quicklinks to frequently cited parts of the 2007 IPCC publication make this site great for everyone, regardless of background.

http://www.globalwarmingart.org provides many clear diagrams depicting the results of climate scientists. The images and captions make the science accessible for college and elementary classrooms alike.

http://www.museumoftheearth.org – The Outreach section of Museum of the Earth's website includes highlights of the many aspects of our Global Change Project (http://www.museumoftheearth.org/outreach.php?page=overview/global change).

http://globalwarming.house.gov explains the current status of federal legislation, as well as media coverage of the debates.

http://geochange.er.usgs.gov is a repository of research on how different regions of the U.S. might be affected by anthropogenic climate change.

http://www.globalchange.gov is another excellent source of regional climate information, highlighting the key issues that will affect each region. It also stores many excellent resources on climate science.

http://www.apolloalliance.org highlights popular news media as they relate to climate and clean energy. It features many stories on alternative energies, legislation, and climate science.

http://www.ncdc.noaa.gov/paleo/ctl explores weather, climate, and climate change using different scales. A great resource for classrooms, with imagery, and an accompanying tutorial.

http://www.eere.energy.gov – The United States Department of Energy website, a source for understanding the aspects of energy efficiency and alternative energy sources. This site contains a "do-it-yourself" energy audit for your home or office.

GLOSSARY

aerosol – a suspension of very fine solid or liquid particles in a gas, like sulphates, in the atmosphere.

albedo – the percent of solar energy that a surface reflects back into space.

anthropogenic – made or caused by humans.

atmosphere – the collection of gases that surround a planet.

biodiversity – the variety of life forms found in a specific environment.

biosphere – all organic matter (plants, animals, people), both living and non-living, on Earth.

calcination – a chemical reaction during the production of concrete that releases CO_2 into the atmosphere.

cap and trade – trading system that uses economic incentives to achieve reductions in the emissions of pollutants; see endnote 5.

carbon – the sixth element on the periodic table; a major component of all living things.

carbon cycle – the exchange and recycling of carbon between the geosphere, hydrosphere, atmosphere, and biosphere.

carbon dioxide (CO_2) – a molecule containe one atom of carbon and two atoms of oxygen, produced especially from the burning of fossil fuels, from animal respiration, and through the recycling of natural carbon sinks, such as limestone dissolution.

carbon neutral – property of an activity, individual, or machine that emits no more carbon into the atmosphere than it absorbs or offsets through the action.

carbon sequestration – the process of capturing CO_2 through biological or physical processes, and removing it from the atmosphere

chaotic – a lack of structure or predictability.

chemical weathering – the breakdown of rocks at the surface of the Earth by chemical change.

climate – the average weather conditions (temperature, rain and snow fall, snow and ice groundcover, and cloud cover) in a geographic region over at least three decades.

climate carbon wedge – a tool for envisioning how high-carbon-emitting energy sources can be replaced by "alternative" sources; see Figure 44.

climate proxy – an alternative to direct measurements of climate variables; data from sources like tree rings, lichens, and pollen are used to infer climate information.

climate system – the interconnected pieces, such as precipitation, wind, ocean currents, and topography, that together influence our climate; altering one aspect of this system impacts others.

climatologist – scientist who specifically works on climate research.

cloud – a collection of water vapor in the atmosphere.

complexity – having multiple facets, for which altering one facet has a complicated and possibly unpredictable effect on facets.

condensation nucleus – suspended particle in the air that can serve as a seed for water molecules to attach to, which is crucial to the formation of clouds.

consensus – general agreement about an issue held by the great majority of specialists who work on that issue.

contrail – abbreviation for "condensation trail"; the condensation of water droplets or ice crystals from the atmosphere, visible in the wake of an airplane, rocket, or missile.

convection current – a vertical circular movement created by warm air/water rising and forcing cool air/water down.

coral – marine, mostly colonial, animals that produce a hard skeleton out of calcium carbonate.

coral bleaching – a symptom of coral sickness whereby the coral become whitened, or bleached, as a result of the disease. This whitening is due to the expulsion of symbiotic algae, called zooxanthellae, from corals' tissue. Many coral sicknesses are brought about by increase in ocean temperature.

cryosphere – that part of the Earth's surface where water exists in solid form; this includes all major forms of ice, including sea ice, glaciers, ice sheets, ice caps, and permafrost.

dendrochronology – the study of climate change as recorded by tree ring growth.

downwelling – the opposite of upwelling, often caused by the sinking of denser (colder or more saline) surface water.

eccentricity – the change in the shape of Earth's orbit on a 100,000-year cycle from a circular to a more elliptical shape.

El Niño – Southern Oscillation (ENSO) represented by fluctuating temperatures and air pressures in the tropical Pacific Ocean. During El Niño, the eastern Pacific experiences warmer water and higher air pressure than the western Pacific, changing rainfall patterns, eastern Pacific upwelling, and weather variables globally. "La Niña" is the reverse. ENSO occurs every 3 to 7 years.

Faint Young Sun Paradox – The Sun was a weaker source of energy earlier in its history, perhaps as much as 30% weaker 4 billion years ago than it is currently. The "paradox" is how the Earth was so warm (with liquid water) with so little energy from the Sun; the solution might be a strong greenhouse atmosphere.

feedback – the response of a system on some variation that, in turn, adds to or removes from the response of the original variation.

forcing – a change that has a directional impact on what is being changed (*e.g.*, a solar forcing on the Earth directly impacts the heat of the Earth in a positive direction).

fossil – the bodily remains or traces of life from the geologic past that are preserved in the Earth's crust.

fossil fuel – a non-renewable, carbon-based fuel source, such as oil, gas, or coal, developed from preserved fossil organisms.

freshening – a decrease in the salt content of ocean water caused by precipitation or melting ice.

geosphere – the solid Earth, from the surface to the core.

glacial interval – cooler period during which ice sheets expanded over the northern parts of the Northern Hemisphere; see also interglacial interval,

glacier – a very large piece of ice that sits at least partly on land and moves under the force of gravity.

Goldilocks Principle – the tenet that states that a parameter must fall within certain margins to be successful, as opposed to reaching extremes.

greenhouse effect – the retention in Earth's atmosphere of some of the heat energy absorbed the Earth that would otherwise escape into space.

greenhouse gas – a gas in Earth's atmosphere that traps energy in the form of heat; carbon dioxide, water vapor, and methane are the most important examples.

heat island – an area with substantially increased heat energy caused by changes in land use; this is usually associated with urban centers.

Heinrich event – a period during the last 100,000 year glacial period in which large numbers of icebergs broke off of glaciers and ice sheets in the Arctic and floated into the northern Atlantic Ocean. Such an event is evidenced by layers of pebbles found in sediment cores from the northern Atlantic, deposited by iceberg melting and dropping of the pebble debris contained within.

hydrosphere – all of the water (including ice) on planet Earth.

hypothesis – an idea to explain an event. Broad explanatory hypotheses (such as gravity, cells, evolution, and plate tectonics) that have been tested and supported by numerous forms of independent evidence, without contradictory evidence, are often called a theory.

ice age – one of several intervals in Earth's history marked by large continental glaciers. Colloquially, this usually refers to the most recent growth of ice sheets, approximately 20,000 years ago, over most of Canada and the northernmost U.S., or to the whole interval since approximately 2.8 million years ago that continental ice sheets have been been growing and shrinking

ice core – a sample of ice collected by drilling into a glacier or ice sheet, in the form of a long, solid cylinder of ice. Chemical data contained within the solid water and air bubbles trapped in the ice provides data on past atmospheric processes.

insolation – the amount of solar radiation received over a particular area (such as a square meter) over a unit of time (such as a day); insolation is average irradiance over a unit of time.

interglacial interval – a period in Earth's history between glacial advances. There have been approximately 50 glacial advances and interglacials in the past 2.8 million years.

IPCC (Intergovernmental Panel on Climate Change) – an international group of climate scientists working to understand climate change and present applicable scenarios to policy makers and the public at large.

irradiance – the amount of solar energy over a unit area at a particular instant, that is, the intensity of sunlight, measured in Watts per square meter.

isotope – an atom of one element that has a different number of neutrons, and therefore a different mass, from another atom of the same element.

Keeling curve – a graph that clearly shows a continuing increase of CO_2 on top of relatively pollution-free Mauna Loa volcano in Hawaii, since before 1960.

lake effect – precipitation (usually snow) caused by the movement of weather systems over a large water body, especially a lake, which picks up water from the lake and deposits it in the form of precipitation across an adjacent land mass.

Last Glacial Maximum – to the most recent expansion of glaciers in the Northern Hemisphere, approximately 20,000 years ago.

leaf margin analysis – a proxy that uses the shape of modern and fossil leaves to help to reconstruct ancient environments and climates.

lithosphere – the rigid outermost layer of the Earth, including the crust and outermost mantle, that moves on weaker underlying layers through the process of plate tectonic movement.

Little Ice Age – a relatively modest cooling period of less than 1°C in the Northern Hemisphere during the 16th to 19th centuries.

mass extinction – event involving extinctions of dramatic numbers of organisms during a relatively short interval of geologic time.

Medieval Warm Period – an historically warm period in Earth's history, approximately 1,000 years ago, during which the Viking people inhabited Greenland.

methane (CH_4) – a greenhouse gas produced from the burning of fossil fuels, as well as through digestion of organic matter by living animals. Methane is present in the atmosphere in lower concentrations than CO_2, but is still a major contributor to the greenhouse effect.

Milankovitch Cycles – periodicities caused by the shape and movement of Earth's orbit, which brings Earth nearer to or farther from the Sun, or changes its tilt, creating periods of higher and lower solar irradiance and absorption; these cycles result in warming and cooling of Earth.

model – a computer generated simulation of a complex pattern, such as the climate system, projected through time.

negative feedback – a change caused by changes already occurring that decreases the rate of change or cancels it out. Note that a decrease in rate does not necessarily mean that the change itself is negative.

obliquity – the change of the angle of Earth's axis, ranging 22-24° from normal, and occuring on a 40,000-year cycle.

ozone (O_3) – a form of molecular oxygen that contains three oxygen molecules. Ozone is harmful to humans and animals when inhaled, and is therefore an air pollutant in the lower atmosphere. In the upper atmosphere, it helps to block the amount of ultraviolet light that reaches the surface of Earth.

paleoclimatology – the study of past climates, commonly using estimates of past climate conditions such as isotopes, tree rings, pollen, and other proxy evidence.

palynology – the study of modern and fossil pollen, spores, and other microscopic plant matter.

parameterization – calculations based on the important parameters (or characteristics) of a given model.

peer-review – process by which scientific data and analyses are reviewed, criticized, and often improved prior to publication.

permafrost – ground that is frozen year-round, common throughout much of the northern reaches of continents in countries such as Canada and Russia.

plate tectonics – the scientific theory that Earth's crust is composed of a series of large and small plates that very slowly move across the surface of the Earth. Plate tectonics is responsible for the distribution of the Earth's continents, for the uplift and position of mountain ranges, and for many other features of the Earth's surface.

positive feedback – a change caused by changes already occurring that increase the pace of change. Note that the increase in pace does not necessarily mean that the change itself is for the better.

precession – the small variation in the direction that Earth's axis points relative to the fixed stars; commonly called "wobble."

prediction – a statement based on scientific observations and data that claims the possibility of some future outcome.

proxy – an alternative to direct measurements of a given variable; data from indirect sources gathered to infer information on what is being measured.

radiative forcing – the net influence that one or more factors, such as greenhouse gases, have on the energy absorbed by the Earth.

rain shadow – an area on one side of a mountain that experiences little rainfall. The mountain blocks the passage of rainclouds to the far side of the mountain, creating a "shadow" area that receives little rain.

rolling average – method to eliminate some of the "noise" from a wildly fluctuating dataset that averages the data from an individual data point with data points before and after it (see Appendix 3).

scale – the depth or magnitude at which something is being examined; either in the amount of time, the amount of space, or both.

science – the process of understanding the natural world through means of observation, hypothesizing, testing, and recording. In science, hypotheses are constantly under scrutiny, in an attempt to falsify information and find better alternative hypotheses.

sea ice – ice that forms at the Earth's surface when seawater freezes.

seafloor spreading – when two adjacent oceanic plates move apart at a mid-ocean ridge (such as the one in the middle of the Atlantic) due to the eruption of lava at the top of the ridge.

sediment core – a sample of sediment collected by drilling into soft sediment or sedimentary rock, yielding a long, solid cylinder.

sunspot – a visual phenomenon appearing as a dark spot on the surface of the Sun, and indicating the presence of a solar flare.

system – a combination of parts that interact with each other to create properties that might not exist or be evident in the individual parts examined separately.

tectonic plate – one of the large or small plates making up the surface of the Earth; see plate tectonics.

temperature anomalies – changes in temperature measured by comparison to a recognized baseline; see Appendix 4.

theory – in science, an idea or set of ideas and hypotheses that connects, explains, and is supported by a large number of observations.

upwelling – the phenomenon that occurs when wind currents push the surface ocean water away from the coast, toward the ocean, causing the cool, deep bottom waters to rise to the surface near the coast.

weather – fluctuations in local conditions of temperature, amount of rainfall or snowfall, snow and ice cover, wind direction and strength, and other factors, which last hours, days, weeks, or months.

Younger Dryas event – a 1,200-year interval of colder temperatures that punctuated a warming trend that began after the Last Glacial Maximum, approximately 13,000 years ago.

APPENDIX 1
THE GEOLOGIC TIME SCALE

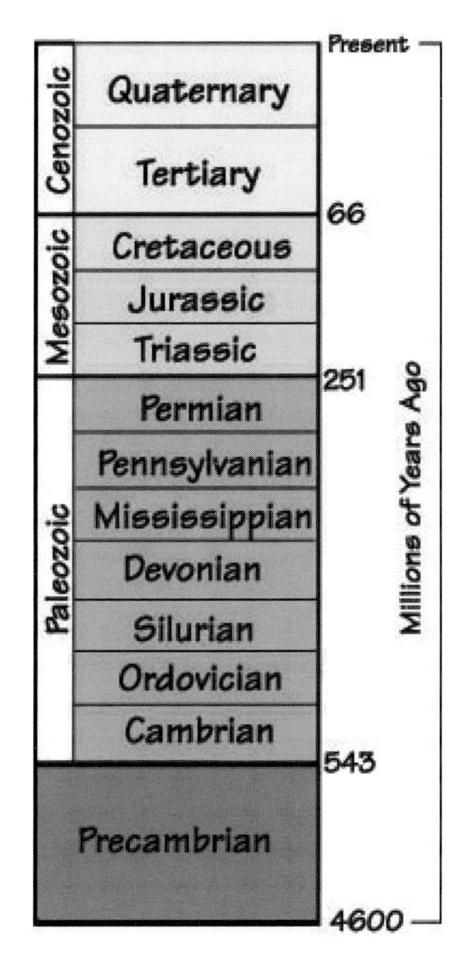

The Geologic Time Scale is the internationally accepted system for telling time in the Earth's history. The names on the Geologic Time Scale are labels for different groups of fossil organisms that lived at different times. The numbers on the Geologic Time Scale are based on using radiometric dating on rocks that occur above or below particular fossils. These dates could change as refinements are made or new rocks are discovered.

APPENDIX 2
USE OF CELSIUS VS. FAHRENHEIT

Celsius and Fahrenheit are two units of measurement for temperature. Fahrenheit is used commonly in the United States; most other countries, however, use Celsius as the standard unit of measurement. A unit itself is an arbitrary reference intended to help us understand the scale of what is being measured. A unit of distance measurement could be a school bus or a pencil. A unit of temperature measurement could be clothes, such as a t-shirt day, a sweater day, or even a two-sweater day. In fact, the band Three Dog Night got their name from an aboriginal tribe in Australia that slept with a dog on cool nights, two dogs on cold nights, and three dogs on very cold nights. These are all units of measurement.

To help standardize units of measurement, the "standard" units were adopted by the U.S., and "metric" units were adopted by most other countries. Metric units, which include Celsius (C), are based on powers of ten, so that 0°C is freezing and 100°C is boiling. In Fahrenheit (F), the freezing point is 32°F and the boiling point is 212°F.

Because powers of ten make calculations easier, and because most countries in the world measure their temperature in Celsius, all of the figures in this book are in Celsius. In the text, both Celsius and Fahrenheit are used, with Celsius first, to remain consistent with figures but also to give readers a sense of temperature using both units, whichever is more familiar to them.

APPENDIX 3
ROLLING AVERAGES

Because there are sometimes so many individual data points in a graph, the "noise," or differences from one point to the next, can hide more relevant long-term trends. As we can see in the graphs below, if the focus is placed on the individual data points, the data seem to show sporadic jumping from one point to another (top graph). But, if we step back from individual data points to look at the graph as a whole, we can see that the data points all trend upward (center graph).

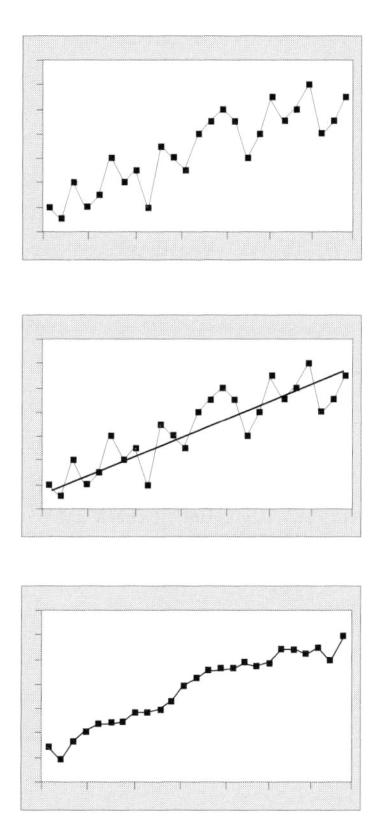

Temperature data over spans of any interval of time from days to millions of years have substantial variation. Sometimes it is hard to see the general climate trend due to the amount of "noise." A rolling average takes the data from an individual data point, and averages it with the data points before and after it, to get rid of some of the "noise." So, in a five-year rolling average of temperature, 2002 global temperature is averaged with that from 2000, 2001, 2003, and 2004, to arrive at a less noisy dataset. The bottom graph is the same data as shown in the above diagrams, but with a rolling average imposed on the data. See how this removes most of the noise and makes the long-term trend more easily seen? Also notice that the beginning and end of the graph still have some "noise." This is because there are no additional data with which to average those data points.

APPENDIX 4
TEMPERATURE ANOMALIES

Most graphs in climate science, and certainly in this book, use **temperature anomalies** instead of true temperatures. Because scientists say that climate is changing, they plot the change in temperature instead of the temperature; the curves look the same, but the changes are measured by comparison to a recognized baseline. This baseline is usually an average global temperature for an extended period of time, such as 1960-1990.

In a temperature anomaly graph, the change in temperature (usually in °C) ranges from something like -3° to +3°. This is the amount of temperature change, positive or negative, that has been or could be seen in the recorded time interval. It is important to remember that the baseline could differ from graph to graph, and so one might show a 2° increase in temperature over 50 years, whereas another might only show a 0.5° increase. This does not mean that there is an argument about how much temperature has changed; it simply means that scientists use the baseline temperature that makes the most sense with their data.

APPENDIX 5
IMPACTS OF CLIMATE CHANGE IN NEW YORK STATE

Because of the many factors influencing future climate change, including future legislation and personal actions, it is impossible to say with certainty what impacts will be felt in New York State (NYS). However, organizations like the EPA and IPCC have been working to establish basic predictions of what different regions can expect as a result of future climate change.

NYS temperature: It is estimated that, by the year 2100, NYS will have experienced an average temperature increase of 2°-4°C (4°-8°F). This is the equivalent of moving from, for example, Syracuse (8.5°C or 47.4°F) to somewhere between Pittsburgh (10.2°C or 50.3°F) and Washington, DC (12.1°C or 53.8°F). This is in addition to the more than the 0.5°C (1°F) average annual rise in temperature that has been felt already in NYS. Temperature is predicted to increase more greatly in summer and fall months, and increase less in winter and spring months.[147]

NYS precipitation: In the last 50 years, average precipitation in NYS has increased by up to 20% in some areas. Yearly precipitation is expected to increase by another 10-20% by 2100 based on IPCC projections. However, this will not be evenly distributed. It is expected that winter precipitation in the northeast will increase by 15-20%, whereas summer precipitation is expected to decrease by 5-10%.

With changes in the seasonal precipitation cycle and increased evaporation, lake levels could lower, impacting marshlands, wildlife, and human recreation. It is expected that more extreme precipitation events will occur as a result of climate change.

Currently very severe thunderstorms and tornadoes are uncommon in the northeastern U.S., and upstate New York usually feels the effects of only the edges of dying hurricanes as they travel over land (although New York City is more likely to receive a big hit along the coast). These events, however, could become part of our landscape rather than the occasional anomaly. Because rapid precipitation results in surface-water runoff, with little rainwater absorbing into the ground, restriction of precipitation to just a few extreme events contributes to more drought-like conditions.

NYS forest impacts: Like all organisms, trees are adapted to a certain set of climate conditions, and NYS has a large amount of forested land. The Adirondack State Park, with 6 million acres of forest, is the largest forested area east of the Mississippi River. It represents an extremely important and significant hardwood ecosystem. As a result of projected climate change, the forests of maple, birch, and beech could die off in NYS and begin thriving farthre north. The extent of forested areas in NYS could experience a decline of 10-25% as a result of climate change.[148]

In addition to loss of habitat, NYS actively harvests maple sap and apples, both of which would be negatively impacted by a warming climate. In the short term, an earlier, warmer spring could lessen the amount of maple sap available during the sugaring season. Maple sap is harvested during a transitional time in early spring, when the weather goes through a freeze-thaw cycle with very cold nights and days above freezing; this cycle encourages movement of sap through the tree. With an earlier and quicker spring, fewer days of harvesting and a diminished harvest are expected. In the long term, across the state, 50-70% of maple forests could be lost by 2100.

Apple trees would also be negatively impacted by climate change. Many apple trees need a certain number of days below freezing to set large amounts of fruit. With warmer winters, many traditional varieties of apples will no longer produce large amounts of big fruit, and with warmer and earlier springs, the trees will bloom earlier. Spring temperatures are incredibly variable, and if a bloom comes early, followed by frost, the apple crop could be significantly damaged.

NYS lake and river systems: It is expected that, by the year 2100, the levels of Lake Ontario and Lake Erie could drop up to 1.5 meters (5 feet) as a result of increased evaporation and decreased recharge in lake basins. Ice coverage on the Great Lakes and other inland lakes has already experienced decline and is expected to continue to decline in future years.

River systems are expected to experience declining fish populations throughout NYS as a result of increased water temperature and turbidity (sedimentation), among other things. These factors are expected to negatively impact important fish, such as trout, in river and lake systems.

NYS coastal areas: New York's coast is currently experiencing a rise of approximately 25 centimeters (10 inches) per 100 years, but it is likely to rise an additional 56 centimeters (22 inches) by 2100. Sea-level rise can erode beaches, erase coastal wetlands, contaminate well water, and increase the chance of major storm-related flooding. New York City could experience storms and hurricanes that would allow seawater to flood Lower Manhattan, southern Brooklyn, and Queens, as well as Staten and Coney islands.

NYS human health: Even with just a 0.5°C (1°F) warming from climate change, heat-related deaths in NYC could rise from 300 to 700 per summer. On the other hand, winter mortality rates are expected to decline as climate warms. Air pollution in NYC, which is caused by increased ground-level ozone (a major component of smog), is expected to rise by 4% if temperature rises the predicted 2°-4°C (4°-8°F).[149]

These health concerns are in addition to the expected habitat expansion that warmer temperatures will create for the disease-carrying insects that flourish in warmer areas and spread diseases like Lyme disease and malaria through human populations. According to the Union of Concerned Scientists, vector-borne diseases like Lyme disease and West Nile encephalitis have expanded widely across the northeast. Although the current increase is blamed on changes in land use, it is a good model of how climate change could impact human health in NYS.[150]

ENDNOTES

[1] In this book, we prefer the term "climate change" over "global warming." Although average global temperature is increasing, dramatically and at an alarming rate, the warming trend is not and has never been predicted to be uniform: some areas will feel more substantial warming than others, and a few regions of the world will even experience cooling. Therefore, "climate change" is a more encompassing phrase for the phenomena that the planet is experiencing.

[2] Good discussions of the broad connections between climate and civilization are provided by: Fagan, Brian M., 2001, *The Little Ice Age: How Climate Made History, 1300-1850*, Basic Books, New York, 272 pp.; and Behringer, Wolfgang, 2009, *A Cultural History of Climate*, Polity Press, Cambridge, U.K., 280 pp.

[3] May, Elizabeth, & Zoe Caron, 2008, *Global Warming for Dummies*, John Wiley & Sons, Somerset, New Jersey, 352 pp.

[4] The book *An Inconvenient Truth: The Planetary Emergency of Global Warming and What We Can Do About It* was published by Rodale Books, Emmaus, Pennsylvania, in 2006. The film *An Inconvenient Truth*, directed by Davis Guggenheim and released by Paramount Classics also in 2006, won Academy Awards for Best Documentary Feature and Best Original Song.

[5] Emissions trading (also known as **cap and trade**) uses economic incentives to achieve reductions in the emissions of pollutants. The government sets a limit, or cap, on the amount of a pollutant that can be emitted. Companies or other groups are issued emission permits and are required to hold an equivalent number of "allowances" (or "credits") that represent the right to emit a specific amount of the pollutant. The total amount of allowances and credits cannot exceed the cap, limiting total emissions to that level. Companies that need to increase their emission allowance must buy credits from those who pollute less. The transfer of allowances is referred to as a "trade." In effect, the buyer is paying a charge for polluting, whereas the seller is being rewarded for having reduced emissions by more than was needed. Thus, in theory, those who can reduce emissions most cheaply will do so, achieving the pollution reduction at the lowest total cost to society. For more details, see http://en.wikipedia.org/wiki/Emissions_trading (last accessed 17 March 2010).

[6] Information on the outcome of the December 2009 Copenhagen conference can be found at http://www.nytimes.com/2010/03/10/science/earth/10climate.html

and http://unfccc.int/2860.php (both last accessed 17 March 2010).

[7] Intergovernmental Panel on Climate Change (IPCC), 2007, *Contributions of Working Groups I, II, and III to the Fourth Assessment Report of the Intergovernmental Panel on Climate Change*, R. K. Pachauri & A. Reisinger , eds., Geneva, Switzerland, 104 pp.

[8] **Climate** is defined by the World Meteorological Organization as the average weather pattern of a region for over 30 years (WMO, *Climate*, http://www.wmo.int/pages/themes/climate/index_en.php, last accessed 17 February 2010).

[9] Rodo, Xavier, & Francisco A. Comin, 2003, *Global Climate: Current Research and Uncertainties in the Climate System*, Springer, New York, 286 pp.; and Johansen, Bruce E., 2008, *Global Warming 101*, Greenwood Press, Westport, Connecticut, 194 pp.

[10] Johansen, 2008 (see endnote 9).

[11] Because it is salty, seawater is denser than freshwater, and so the density difference for seawater with ice is greater than that for freshwater and ice.

[12] National Oceanic and Atmospheric Administration (NOAA), *National Weather Service*, http://www.nws.noaa.gov (last accessed 17 February 2010).

[13] All light consists of waves. The distance between the peaks of these waves is their wavelength. Sunlight includes waves of a variety – or spectrum – of wavelengths. At one end of the spectrum, shorter wavelengths include blue light and its invisible neighbor, ultraviolet light. At the other end, longer wavelengths include red light and its invisible neighbor, infrared radiation, which we feel as heat.

[14] IPCC, 2007 (see endnote 7).

[15] Johansen, Bruce E., 2002, *The Global Warming Desk Reference*, Greenwood Press, Westport, Connecticut, 353 pp.

[16] Pickering, Kevin T., & Lewis S. Owen, 1997, *An Introduction to Global Environmental Issues, 2nd ed.*, Routledge, New York, 512 pp.

[17] Johansen, 2008 (see endnote 9).

[18] Good overviews of plate tectonic activity and climate change in Earth history can be found in two excellent college-level textbooks: Stanley, S. M., 1999, *Earth System History*, W. H. Freeman, New York, 615 pp.; and Ruddiman, W., 2008, *Earth's Climate: Past and Future, 2nd ed.*, W. H. Freeman and Company, New York, 388 pp.

[19] Variation in incoming solar radiation is due, in part, to Milankovitch Cycles, but the changing shape of Earth's orbit (**eccentricity**) is caused by the gravitational forces of other planetary bodies, like Jupiter and Saturn, which are orbiting the Sun as well.

[20] A cooler climate of several thousand years or less during an interglacial period is technically called a **stadial** (the Younger Dryas is one example), and an analogous warmer climate during a glacial interval is called an **interstadial**.

[21] **El Niño** is formally known as the El Niño-Southern Oscillation (ENSO). It is defined as warming of at least 0.5°C (0.9°F) in ocean-surface temperatures compared with the average value for at least five months. The causes of El Niño are not well-understood, but its results are well known and often extremely serious. When the warm water spreads from the western Pacific and the Indian Ocean to the eastern Pacific, it brings rain with it, usually causing extensive drought in the western Pacific and higher rainfall in the eastern Pacific and even into southern North

America. The warm current of nutrient-poor tropical water, heated by its eastward passage in the Equatorial Current, replaces the cold, nutrient-rich surface water of the Humboldt Current off of the western coast of South America. This causes a decline in fish stocks and also of populations of the animals that eat fish, such as marine birds and mammals. More information on El Niño can be found at http://www.ncdc.noaa.gov/teleconnections/enso and http://oceanworld.tamu.edu/resources/oceanography-book/heatbudgets.htm (last accessed 17 March 2010).

[22] The relationship between El Niño and global warming is discussed by Fagen, Brian, 2000, *Floods, Famine, and Emperors: El Niño and the Fate of Civilizations,* Pimlico, London, 304 pp.; and at http://www.realclimate.org/index.php/archives/2006/05/el-nino-global-warming (last accessed 17 March 2010).

[23] Heat is radiated from Earth's core as a product of the decay of radioactive material. This heat drives convection cells within the mantle of Earth which, in turn, cause the movement of the plates along the outer surface of the mantle.

[24] Pickering & Owen, 1997 (see endnote 18).

[25] It is estimated that energy from the Sun was approximately 30% lower 4 billlion years ago, shortly after the Earth was formed (Cronin, Thomas M., 2009, *Paleoclimates: Understanding Climate Change Past and Present,* Columbia University Press, New York, 441 pp.).

[26] Stable isotope composition can be measured either from air bubbles trapped in ice or from the ice matrix itself (Cronin, 2009, see endnote 25).

[27] A good general discussion of ice cores and their study is provided by: Alley, Richard B., 2002, *The Two-Mile Time Machine: Ice Cores, Abrupt Climate Change, and Our Future,* Princeton University Press, Princeton, New Jersey, 240 pp.

[28] **Ice age** is a popular but not very precise term for a geologically recent time interval during which glaciers expanded over the continents in the Northern Hemisphere. Use of this term, however, can be confusing because it can refer to a longer interval and/or a shorter piece of it. The longer interval began approximately 3 million years ago, during which glaciers expanded and contracted in the Northern Hemisphere, and prior to which there had been little or no ice in the northern polar region. During this longer interval, expansions of the ice are called **glacials** and retreats of the ice are called **interglacials**. Glacials and interglacials have alternated over the past 3 million years on a cycle of roughly 100,000 years, due mostly to the effect of **Milankovitch Cycles** in the Earth's orbit on solar insolation. The term "ice age" can also refer to a much shorter interval; the most extreme extent of the most recent of these glacial intervals was approximately 20,000 years ago (also called the **Last Glacial Maximum**). In this book, we use "ice age" to represent the last 3 million years of Earth's history, and "Last Glacial Maximum" or "most recent glaciation" to represent the most recent glacial interval.

[29] The shift from 40,000-year cycles to 100,000-year cycles, which occurred approximately 700,000 years ago, is known as the "Mid-Pleistocene Transition." There has been considerable research in the past decade on the cause of this transition; explanations have included changes in amplification of Milankovitch insolation cycles involving weathering and erosion, CO_2 levels, Southern Ocean ice cover, tropical climate dynamics, and other factors (Cronin, 2009; see endnote 25).

[30] See discussion in Chapter 10, FAQ 16, for more information.

[31] Modified after a graph on page 58 of Broecker, Wallace S., 2003, *Fossil Fuel, CO2*

and the Angry Climate Beast, Eldigio Press, Palisades, New York, 112 pp.

[32] Modified after a graph presented by Hansen, James, Makiko Sato, Reto Ruedy, Ken Lo, David W. Lea, & Martin Medina-Elizade, 2006, Global temperature Change, *Proceedings of the National Academy of Sciences of the United States of America,* 103(39): 14288-14293.

[33] IPCC, 2007 (see endnote 7).

[34] IPCC, 2007 (see endnote 7).

[35] IPCC, 2007 (see endnote 7).

[36] IPCC, 2007 (see endnote 7).

[37] IPCC, 2007 (see endnote 7).

[38] IPCC, 2007 (see endnote 7).

[39] IPCC, 2007 (see endnote 7).

[40] IPCC, 2007 (see endnote 7).

[41] Stora Enso report (undated) *Curbing Climate Change*, http://www.storaenso.com/sustainability/publications/Sustainabilitybooklets/Documents/40075_SE_Booklet_ccc_LR.pdf (last accessed 21 March 2010).

[42] Actually Arrhenius thought that anthropogenic warming of the globe would be a good thing, because it would allow people to live more easily at higher latitudes and would increase agricultural growing seasons. See Tennesen, Michael, 2008, *A Complete Idiot's Guide to Global Warming, 2nd ed.,* Alpha Books/Penguin Group, New York, 352 pp.

[43] Intergovernmental Panel on Climate Change (IPCC), 1995, *Second Assessment Synthesis of Scientific-Technical Information Relevant to Interpreting Article 2 of the UNFCCC,* Geneva, Switzerland, 64 pp.

[44] Intergovernmental Panel on Climate Change (IPCC), 2001, *Climate Change 2001: Synthesis Report. A Contribution of Working Groups I, II, and III to the Third Assessment Report of the Intergovernmental Panel on Climate Change*, R. T. Watson & the Core Writing Team, eds., Cambridge University Press, Cambridge, United Kingdom and New York, 398 pp.

[45] IPCC, 2007 (see endnote 7).

[46] Pages 23-24 in Emanuel, Kerry, 2007, *What We Know About Climate Change*, Massachusetts Institute of Technology Press, Cambridge, Massachusetts, 85 pp.

[47] A "natural balance" is not actually exact, which is why there is change over geologic time. Before humans became such a large influence, however, carbon sources and sinks were very close to being balanced on human timescales.

[48] IPCC, 2007 (see endnote 7).

[49] IPCC, 2007 (see endnote 7).

[50] IPCC, 2007 (see endnote 7).

[51] IPCC, 2007 (see endnote 7).

[52] Other carbon sinks caused by vegetation include peat bogs, marshes, and oceans. These are all in relative equilibrium unless deforestation, loss of habitat, or some other ecological imbalance causes a rapid increase in the death of the vegetation compared to background rates.

[53] Pickering & Owen, 1997 (see endnote 16).

[54] Cronin, 2009 (see endnote 25).

[55] For further discussion of climate models, see Emanuel, 2007: 39 (see endnote 46); chapter 5 of Houghton, John, 2009, *Global warming: The Complete Briefing,*

4th ed., Cambridge University Press, Cambridge, U.K., 438 pp.; and http://www.realclimate.org/index.php/archives/2005/01/is-climate-modelling-science (last accessed 17 March 2010).

[56] From the Geophysical Fluid Dynamic Laboratory in Princeton, New Jersey; for further information, see http://www.gfdl.noaa.gov (last accessed 01 April 2010).

[57] **Upwelling** occurs when wind currents push the surface ocean water away from the coast, toward the ocean, causing the cool, deep bottom waters to rise to the surface near the coast. **Downwelling** is the reverse phenomenon, caused by the sinking of denser (colder or more saline) surface water.

[58] IPCC, 2007 (see endnote 7).

[59] A number of studies published since the 2007 IPCC report suggest that this might be a conservative estimate. According to this research, even if greenhouse gas emissions were to stabilize at present levels immediately, global temperatures will still rise by approximately 2.4°C by 2100, and that it could take as long as 1,000 years beyond this for temperatures to return to present levels. For details see Greene, C. H., D. J. Baker, & D. H. Miller, 2010, A very inconvenient truth, *Oceanography*, 23(1): 214-218.

[60] For more details on "tipping points" in climate change, see Lenton, T. M., *et al.*, 2007, Tipping elements in the Earth's climate system, *Proceedings of the National Academy of Sciences of the USA,* 105: 1786-1793. A good general discussion of "tipping points" is presented by Pearce, Fred, 2007, *With Speed and Violence: Why Scientists Fear Tipping Points in Climate Change*, Beacon Press, Boston, 278 pp.

[61] On 31 May 2007, NASA Administrator Michael Griffin said in an interview with National Public Radio that he was not sure that global warming was a problem. "I have no doubt that ... a trend of global warming exists ... I am not sure that it is fair to say that is a problem we must wrestle with ... I guess I would ask which human beings, where and when, are to be accorded the privilege of deciding that this particular climate that we have right here today, right now, is the best climate for all other human beings. I think that's a rather arrogant position for people to take." A few days later, Griffin said that he regretted the earlier remarks, saying that "unfortunately, this is an issue that has become far more political than technical, and it would have been well for me to have stayed out of it." See http://www.npr.org/templates/story/story.php?storyId=10571499 (last accessed 17 March 2010).

[62] The 2007 IPCC report (see endnote 7) contained at least three serious overstatements of the harmful consequences of global climate change. These statements, however, had no bearing on the central conclusion that human emissions of greenhouse gases are primarily responsible for climate change. In March 2010, the United Nations appointed a special committee to investigate the internal processes of the IPCC that allowed the statements to escape the otherwise rigorous review to which the report was subjected. The review is expected to recommend stricter checking of sources and more careful wording to reflect uncertainties. (The IPCC's current statement of procedures for preparation and review of its reports is available at http://web.archive.org/web/19960101000000-20090917211948/http://www.ipcc.ch/about/app-a.pdf, last accessed 17 March 2010.)

The first error was a claim that all Himalayan glaciers would likely disappear by 2035. Most scientists, however, believe that it would take another 300 years for the glaciers to melt at the present rate. The report also claimed that global warming could cut rain-fed North African crop production by up to 50% by 2020. A senior IPCC contributor subsequently admitted that there is no evidence to support this claim. The third (and perhaps the most embarrassing) error was the statement that more than half of The Netherlands is below sea level, whereas the correct figure is only 26%. (From *The Times*, London, 10 March 2010, http://www.timesonline.co.uk/tol/news/world/us_and_americas/article7055999.ece, last accessed 19 March 2010).

[63] IPCC, 2007 (see endnote 7).

[64] Intergovernmental Panel on Climate Change (IPCC), 1990, *First Assessment Overview and Policymaker Summaries and 1992 IPCC Supplement*, Geneva, Switzerland, 168 pp.

[65] IPCC, 2007 (see endnote 7).

[66] Mooney, Chris, 2007, *Storm World: Hurricanes, Politics, and the Battle Over Global Warming*, Houghton Mifflin Harcourt, New York, 400 pp.

[67] IPCC, 2001 (see endnote 44).

[68] See http://www.nytimes.com/2007/03/03/us/03maple.html, last accessed 17 March 2010.

[69] McMichael, A. J., *et al.,* 2003, *Climate Change and Human Health. Risks and Responses,* World Health Organization, Geneva, 250 pp.; and http://www.who.int/globalchange/en (last accessed 30 March 2010).

[70] Environmental Protection Agency, *Ozone*, http://www.epa.gov/ozone (last accessed February 17, 2010).

[71] Many corals, especially those that build reefs, contain symbiotic single-celled algae that use sunlight to produce food by photosynthesis and contribute part of it to the coral. It is the expulsion of these greenish-brown algae that causes the whitening of the coral tissue known as "bleaching."

[72] Recent studies include: Graham, N. A. J., *et al.,* 2006, Dynamic fragility of oceanic coral reef ecosystems, *Proceedings of the National Academy of Sciences of the United States of America*, 103: 8425-8429; Donner, S. D., 2007, Model-based assessment of the role of human-induced climate change in the 2005 Caribbean coral bleaching event, *Proceedings of the National Academy of Sciences of the United States of America*, 104(13): 5483-5488; Hoegh-Guldberg, O., *et al.,* 2007, Coral reefs under rapid climate change and ocean acidification, *Science*, 318: 1737-1742; Anthony, K. R. N., *et al.,* 2008, Ocean acidification causes bleaching and productivity loss in coral reef builders, *Proceedings of the National Academy of Sciences of the United States of America*, 105(45): 17442-17446; and Carpenter, K. E., *et al.,* 2008, One-third of reef-building corals face elevated extinction risk from climate change and local impacts, *Science*, 321: 560-563.

[73] Acidity is measured by a numerical scale known as **pH**, which expresses the approximate concentration of hydrogen atoms dissolved in a liquid, on a scale of 0 to 14. Pure water is said to be neutral, with a pH of 7.0. Solutions with a pH of less than 7.0 are said to be acidic, and those with a pH greater than 7.0 are said to be basic, or alkaline. Car battery acid has a pH of just under 0, lemon juice is approximately 2, vinegar approximately 4, and milk of magnesia (magnesium hydroxide) approximately 10.5.

[74] Wootton, J.Timothy, Catherine A. Pfister, & James D. Forester, 2008, Dynamic patterns and ecological impacts of declining ocean pH in a high-resolution multi-

year dataset, *Proceedings of the National Academy of Sciences of the United States of America*, 105(48): 18848-18853.

[75] IPCC, 2007 (see endnote 7); and Hoegh-Guldberg *et al.*, 2007 (see endnote 72).

[76] Vermeij, G., 2009, Seven variations on a recent theme of conservation, Pages 167-175, in: *Conservation Paleobiology: Using the Past to Manage for the Future*, G. Dietl & K. Flessa, eds., The Paleontological Society Special Paper 15.

[77] Ruttiman, Jacqueline, 2006, Oceanography: sick seas. *Nature*, 442: 978-980.

[78] Ries, Justin B., Anne L. Cohen, & Daniel C. McCorkle, 2009, Marine calcifiers exhibit mixed responses to CO_2-induced ocean acidification, *Geology*, 37(12): 1131-1134.

[79] See, *e.g.*, Knoll, Andrew A., R. K. Bambach, J. L. Payne, S. Pruss, & W. W. Fischer, 2007, Paleophysiology and end-Permian mass extinction, *Earth and Planetary Science Letters*, 256: 295-313; and Veron, J. E. N., 2008, Mass extinctions and ocean acidification: biological constraints on geological dilemmas, *Coral Reefs*, 27: 459-472.

[80] Good overviews of mass extinction in the geological record are provided by: Erwin, D.H., 2006, *Extinction: How Life on Earth Nearly Ended 250 Million Years Ago*, Princeton University Press, Princeton, New Jersey, 320 pp.; and Hallam, A., 2004, *Catastrophes and Lesser Calamities: The Causes of Mass Extinctions*. Oxford University Press, Oxford, U.K., 288 pp.

[81] Discussion of the value of biodiversity for human well-being is presented by Myers, Norman, 1984, *The Primary Source: Tropical Forests and Our Future*, W. W. Norton, New York, 399 pp.; Beattie, Andrew, & Paul Ehrlich, 2004, *Wild Solutions: How Biodiversity is Money in the Bank, 2nd ed.*, Yale University Press, New Haven, Connecticut, 288 pp.; and Chivian, Eric, & Aaron Bernstein, eds., 2008, *Sustaining Life: How Human Health Depends on Biodiversity,* Oxford University Press, Oxford, U.K., 568 pp.

[82] See http://maps.grida.no (last accessed 21 March 2010).

[83] Discussion of the possible equivalence of current human-caused extinctions with ancient mass extinction is presented by Ward, P. D., 1994, *The End of Evolution: On Mass Extinctions and the Preservation of Biodiversity,* Bantam Books, New York, 301 pp.; Leakey, R., & R. Lewin, 1995, *The Sixth Extinction: Patterns of Life and the Future of Humankind,* Doubleday, New York, 271 pp.; and Glavin, T., 2007, *The Sixth Extinction: Journey Among the Lost and Left Behind,* Thomas Dunne Books/St. Martin's Press, New York, 318 pp.

[84] The actual and potential effects of current and future climate change on biodiversity is a large and rapidly growing subject. Good introductions are presented by Lovejoy, T. E., & L. Hannah, eds., 2005, *Climate Change and Biodiversity,* Yale University Press, New Haven, Connecticut, 418 pp.; and Lawler, J. J., *et al.*, 2009, Projected climate-induced faunal change in the Western Hemisphere, *Ecology*, 90(3): 588-597.

[85] Harvell, C. Drew, *et al.*, 2002, Climate warming and disease risks for terrestrial and marine biota, *Science*, 296: 2158-2162.

[86] These include the American Association for the Advancement of Science, the Union of Concerned Scientists, the American Meteorological Society, the American Geophysical Union, reports from the National Academy of Sciences and the U.S. government's Global Change Research Program, and many others.

[87] In January 2009, a Gallup poll of 3,146 Earth scientists found that 82% answered positively to the question "Do you think human activity is a significant contributing factor in changing mean global temperatures?" Of the 77 climatologists actively engaged in research, 75 (97.4%) answered yes. This supports the results of a 2004 study by Naomi Oreskes [Behind the ivory tower: the scientific consensus on climate change, *Science*, 306(5702): 1686] that looked at the abstracts of nearly 1,000 scientific papers containing the term "global climate change" published during the previous decade. Not one paper rejected the consensus position.

[88] Kennedy, D., 2001, An unfortunate U-turn on carbon, *Science*, 291: 2515.

Strangely, television weather forecasters show much lower levels of acceptance of the consensus on climate change. See Kaufman, Leslie, 2010, Among weathercasters, doubts on warming, *The New York Times*, 30 March 2010, http://www.nytimes.com/2010/03/30/science/earth/30warming.html?th&emc=th (last accessed 30 March 2010) and Homans, Charles, 2010, Hot air: why don't TV weathermen believe in climate change? *Columbia Journalism Review*, January-February 2010, http://www.cjr.org/cover_story/hot_air.php (last accessed 30 March 2010).

[89] IPCC, 2007 (see endnote 7).

[90] Good discussions of the lack of public understanding of and/or engagement with the issue of climate change are provided by: Corbett, J. B., & J. L. Durfee, 2004, Testing public (un)certainty of science: media representations of global warming, *Science Communication*, 26(2): 129-151; Hulme, M., 2009, *Why We Disagree About Climate Change*, Cambridge University Press, Cambridge, U.K., 392 pp.; and Norgaard, K. M., 2009, Cognitive and behavioral challenges in responding to climate change, Background Paper to the 2010 World Development Report, *Policy Research Working Paper 494*, The World Bank, Washington, DC, 74 pp.

[91] Research in the fields of education and psychology suggests that society as a whole lacks the necessary understanding of the nature of science to make policy decisions based on scientific complexity and inherent "uncertainty" (Bradshaw, G. A., & J. G. Borchers, 2000, Uncertainty as information: narrowing the science-policy gap, *Ecology and Society*, 4(1): 7-14).

[92] This point was explored in detail by Freudenburg *et al.* [Freudenburg, William R., Robert Gramling, & Debra J. Davidson, 2008, Scientific certainty argumentation methods (SCAMs): science and the politics of doubt, *Sociological Inquiry*, 78(1): 2-38], who concluded: "Given that most scientific findings are inherently proababilistic and ambiguous, if agencies can be prevented from imposing any regulations until they are unambiguously 'justified,' most regulations can be defeated or postponed, often for decades, allowing profitable but potentially risky activities to continue unabated." A less technical, albeit more partisan, discussion is provided by Mooney, C., 2005, *The Republican War on Science*, Basic Books, New York, 342 pp.; and by Mooney, 2007 (see endnote 66).

[93] The subject of media coverage of climate change is well discussed by Ward, Bud, 2008, *Communicating on Climate Change: an Essential Resource for Journalists, Scientists, and Educators*, Metcalf Institute for Marine and Environmental Report-ing, Narragansett, Rhode Island, 74 pp.; and Mooney, 2007 (see endnote 66).

[94] Ward, 2008 (see endnote 93).

[95] For more discussion of this point, see http://scienceblogs.com/illconsidered/2006/02/temperature-record-reliability-attack.php (last accessed 17 March

2010).

[96] See also http://www.realclimate.org/index.php/archives/2004/12/index/ #Responses (last accessed 17 March 2010).

[97] In late November 2009, more than 1,000 email messages between scientists at the Climate Research Unit of the University of East Anglia in England and other climate scientists around the world were stolen and made public by a still-unidentified computer hacker. Climate-change skeptics have claimed that the messages show scientific misconduct that amounts to the complete fabrication of anthropogenic global warming, and that they constitute a major academic scandal which some have labeled "Climategate." The facts, however, do not support the skeptics' interpretation. The emails, which span a 13-year period, do include some unguarded language, but claims of intentional data manipulation and exclusion of climate skeptics from the scientific literature are completely overblown. A panel of British lawmakers reviewing the case in March 2010 found no evidence to support the skeptics allegations, although it faulted the scientists for expending more effort worrying about how to handle critics than publishing their data. An excellent analysis of the controversy, including references to other sources, is presented at http://www.factcheck.org/2009/12/climategate (last accessed 30 March 2010). In any case, nothing in the emails negates the overwhelming data summarized in this book and elsewhere that global temperatures have been rising and that human burning of fossil fuels is the most likely cause.

[98] Poll by ABC (American Broadcasting Company) News, Planet Green, and Stanford University conducted 23-28 July 2008 (n = 1,000 adults), http://www.pollingreport.com/enviro.htm (last accessed 2 February 2010).

[99] ABC, Planet Green, and Stanford University poll (see endnote 98).

[100] Poll by Cable News Network (CNN) and the Opinion Research Corporation conducted 26-29 June 2008 (n = 1,026 adults), http://www.pollingreport.com/enviro.htm (last accessed 2 February 2010); also Gallup poll conducted 6-9 March 2008 (n = 1,012 adults), http://www.polling report.com/enviro.htm (last accessed 2 February 2010).

[101] ABC, Planet Green, and Stanford University poll (see endnote 98); and CNN opinion research poll (see endnote 100).

[102] Gallup poll conducted 11-14 March 2007 (n = 1,009 adults), http://www.polling report.com/enviro.htm (last accessed 2 February 2010); see also polling results referenced in endnote 97.

[103] Gallup poll conducted 11-13 December 2009, including comparison answers from as early as 2001 (n = 1,025 adults), http://www.pollingreport.com/enviro.htm (last accessed 2 February 2010).

[104] Pew Research Center survey conducted 30 September – 4 October 2009 (n = 1,500 adults), http://www.pollingreport.com/enviro.htm (last accessed 2 February 2010); Yale poll survey conducted 24 December 2009 – 03 January 2010 (n = 1,001 adults); and Leiserowitz, A., E. Maibach, & C. Roser-Renouf, 2010, *Global Warming's Six Americas, January 2010*, Yale University and George Mason University, New Haven, Connecticut, Yale Project On Climate Change, 23 pp.; http://environment.yale.edu/uploads/SixAmericasJan2010.pdf, last accessed 01 April 2010.

[105] See endnote 97.

[106] See endnote 62.

[107] A good discussion of this topic is provided by Hulme (2009; see endnote 90).

[108] Sociologist Kari Marie Norgaard has suggested that citizens of wealthy nations, whose economic well-being depends upon production and/or use of fossil fuels, can show "socially organized denial"" of apparent realities that they just do not want to know about. See Norgaard, K. M., 2006, "We don't really want to know": environmental justice and socially organized denial of global warming in Norway, *Organization and Environment*, 19: 347.

[109] Sagan, C., 1996, *The Demon-haunted World: Science as a Candle in the Dark*, Random House, New York, 457 pp.

[110] "Science and knowledge are intrinsically uncertain, with new information continually altering our perceptions and beliefs ... Science, with its large, complex simulation models of possibly chaotic systems may never produce the needed levels of certainty (to please policy-makers)" (Bradshaw & Borchers, 2000, see endnote 91).

[111] See carbon footprint calculator examples at http://www.nature.org/initiatives/climatechange/calculator, http://coolclimate.berkeley.edu, and http://www.footprint-network.org/en/index.php/GFN/page/calculators (last accessed 17 March 2010).

[112] Based on 1970-2008, U.S. Energy Information Administration, *Annual Energy Review 2008*, published 26 June 2009.

[113] U.S. Energy Information Administration, *Energy and the Environment Explained*, http://tonto.eia.doe.gov/energyexplained/index.cfm (last accessed 17 February 2010).

[114] Energy Star is an international standard for energy efficient consumer products, first created as a U.S. government program by the Clinton Administration in 1992. Australia, Canada, Japan, New Zealand, Taiwan, and the European Union have since adopted the program. Devices carrying the Energy Star logo, especially kitchen and other household appliances, generally use 20-30% less energy than required by federal standards. See http://www.energystar.gov (last accessed 20 March 2010).

[115] The National Action Plan for Energy Efficiency, which operates through the Environmental Protection Agency, discusses the costs and benefits of increasing requirements for energy efficiency in building codes. See their factsheet, *Building Codes for Energy Efficiency*, at http://www.epa.gov/cleanrgy/documents/buildingcodesfactsheet.pdf (last accessed 10 March 2010).

[116] U.S. Energy Information Administration (see endnote 113).

[117] Transportation Research Board, http://www.trb.org (last accessed 17 February 2010).

[118] U.S. Energy Information Administration, 2007 data on total energy consumption per capita (in BTUs); http://tonto.eia.doe.gov/state/state_energy_rankings.cfm?keyid=60&orderid=1 (last accessed 21 March 2010).

[119] Assuming 10,000 miles driven annually, increasing gas mileage from 25 to 35 miles per gallon and 20 pounds of CO_2 emitted per gallon of gas.

[120] Transportation Research Board (see endnote 117).

[121] Transportation Research Board (see endnote 117); and U.S. Department of Energy, *Why is Fuel Economy Important?*, http://fuel-economy.gov/feg/why.shtml (last

accessed 17 February 2010).

[122] Transportation Research Board (see endnote 117).

[123] The BTU (British Thermal Unit) is the traditional unit of energy equal to approximately 1.06 kilojoules. It is the amount of energy needed to heat one pound of water 1°F.

[124] U.S. Energy Information Administration (see endnote 113).

[125] Environmental Protection Agency, *Clean Energy*, http://www.epa.gov/RDEE/index.html (last accessed 2 March 2010).

[126] Based on 1960-2008, U.S. Energy Information Administration, *Annual Energy Review 2008*, published 26 June 2009.

[127] Modified after International Energy Agency, *World Energy Outlook 2007*, 600 http://www.iea.org/textbase/nppdf/free/2007/weo_2007.pdf (last accessed 22 March 2010); and U.S. Energy Information Administration, 2009, *Emissions of Greenhouse Gases in the United States 2008*, Report #DOE/EIA-0573, ftp://ftp.eia.doe.gov/pub/oiaf/1605/cdrom/pdf/ggrpt/057308.pdf (last accessed 17 March 2010).

[128] Solocow, R., & S. Pacala, 2004, Stabilization wedges: solving the climate problem for the next 50 years with current technologies, *Science*, 305: 968-972.

[129] American Wind Energy Association, *Resources*, http://awea.org/faq (last accessed 17 February 2010).

[130] Union of Concerned Scientists, *Clean Energy*, http://www.ucsusa.org/clean_energy (last accessed 4 March 2010).

[131] Nuclear Energy Institute, *Key Issues*, http://www.nei.org/keyissues (last accessed 17 February 2010).

[132] Geothermal Energy Association, *Geothermal Basics*, http://www.geo-energy.org/basics.aspx (last accessed 10 February 2010).

[133] U.S. Department of Energy, *Energy Efficiency and Renewable Energy: Geothermal Technologies Program*, http://www.eere.energy.gov (last accessed 17 February 2010).

[134] Pickering & Owen, 1997 (see endnote 16).

[135] For more on carbon sequestration, see U.S. Department of Energy, *Carbon Sequestration*, http://www.energy.gov/sciencetech/carbonsequestration.htm (last accessed 22 March 2010); and Benson, S. M., & D. R. Cole, 2008, CO_2 sequestration in deep sedimentary formations, *Elements*, 4(5): 325-331.

Carbon sequestration is one example of what is increasingly being called "geoengineering" – attempts to directly mitigate the effects of human-caused environmental change rather than, or in addition to, addressing its causes. Some scientists are now suggesting that reduction of greenhouse gas emissions, no matter how much or how fast, will not be enough to avoid "dangerous" effects of climate change, and "geoengineering will be required." For more details see Greene *et al.*, 2010 (see endnote 59).

[136] Nuzzo, Regina, 2005, Profile of Stephen H. Schneider, *Proceedings of the National Academy of Sciences of the United States of America*, 102(44): 15725-15727.

[137] Peterson, Thomas C., & Michael J. Manton, 2008, Monitoring changes in climate extremes, *Bulletin of the American Meteorological Society*, 89(9): 1266-1271.

[138] Peterson & Manton, 2008 (see endnote 137).

[139] IPCC, 2007 (see endnote 7).

[141] IPCC, 2001 (see endnote 44); and Mann, Michael E., Raymond S. Bradley, & Malcolm K. Hughes, 1998, Global-scale temperature patterns and climate forcings over the past six centuries, *Nature*, 392: 779-787.

[142] National Academy of Science, 2006, *Surface Temperature Reconstructions for the Last 2000 Years*, http://nap.edu/catalog.pho?record_id=11676 (last accessed 1 March 2010).

[143] IPCC, 2007 (see endnote 7).

[144] See endnotes 95 and 135.

[145] See http://geochange.er.usgs.gov (last accessed 17 March 2010).

[146] See endnote 4.

[147] Environmental Protection Agency, http://www.epa.gov (last accessed 4 February 2010); and IPCC, 2007 (see endnote 7).

[148] Environmental Protection Agency (see endnote 147).

[149] Environmental Protection Agency (see endnote 147).

[150] Union of Concerned Scientists, http://www.ucsusa.org/clean_energy (last accessed 4 March 2010).

About the Authors

Warren D. Allmon *is the Director of the Paleontological Research Institution and the Hunter R. Rawlings III Professor of Paleontology in the Department of Earth & Atmospheric Sciences at Cornell University. He received his undergraduate degree from Dartmouth College and his Ph.D. from Harvard University. His research focuses on the ecological and environmental context of evolutionary change, especially in fossil marine mollusks.*

Trisha A. Smrecak *is the Global Change and Evolution Projects Manager at the Paleontological Research Institution. She received her undergraduate degree from St. Lawrence University and her M.S. from the University of Cincinnati. Her geological research focuses on changes in organism-organism relationships with environment. At PRI, she coordinates National Science Foundation grants teaching evolutionary concepts to fifth-to-ninth-grade classrooms and the effects of a changing climate on local and regional environments in concert with New York State 4-H clubs.*

Robert M. Ross *is Associate Director for Outreach at the Paleontological Research Institution and an adjunct professor in the Department of Earth & Atmospheric Sciences at Cornell University. He received his undergraduate degree from Case Western Reserve University and his Ph.D. from Harvard University. His geological research has focused on paleoclimatology and its evolutionary applications. In addition to managing all of PRI's educational programs, he is also active in national pre-college Earth science education reform efforts.*